U0201202

古代佩饰

ACCESSORIES WORE BY ANCIENT CHINESE

“符号中国”编写组◎编著

中央民族大学出版社
China Minzu University Press

图书在版编目(CIP)数据

古代佩饰：汉文、英文 / “符号中国”编写组编著. — 北京：中央民族大学出版社, 2024.3

（符号中国）

ISBN 978-7-5660-2329-2

Ⅰ. ①古… Ⅱ. ①符… Ⅲ. ①服饰文化—介绍—中国—古代—汉、英 Ⅳ. ①TS941.742.2

中国国家版本馆CIP数据核字（2024）第017464号

符号中国：古代佩饰 ACCESSORIES WORE BY ANCIENT CHINESE

编　　著　“符号中国”编写组
策划编辑　沙　平
责任编辑　周雅丽
英文指导　李瑞清
英文编辑　邱　械
美术编辑　曹　娜　郑亚超　洪　涛
出版发行　中央民族大学出版社
北京市海淀区中关村南大街27号　　邮编：100081
电话：（010）68472815（发行部）　传真：（010）68933757（发行部）
（010）68932218（总编室）　　（010）68932447（办公室）
经 销 者　全国各地新华书店
印 刷 厂　北京兴星伟业印刷有限公司
开　　本　787 mm×1092 mm　1/16　印张：10.375
字　　数　134千字
版　　次　2024年3月第1版　2024年3月第1次印刷
书　　号　ISBN 978-7-5660-2329-2
定　　价　58.00元

“符号中国”丛书编委会

唐兰东　巴哈提　杨国华　孟靖朝　赵秀琴

本册编写者

戚琳琳

古代佩饰是指佩戴在人体各部位的饰物，主要包括佩件和首饰两大类。中国的古代佩饰是衣着服饰制度的一个重要组成部分，除了具有美化功能外，还具有一些吉祥寓意、宗教寓意及权、礼观念上的特别意义。

中国古代的佩饰品种繁多、造型各异、工艺精湛，一些饰品虽

People in ancient times wore accessories and jewels on different parts of the body. China's ancient accessories have constituted an essential part of the institution of ancient costumes. Ancient people not only bedecked themselves with accessories but also endowed them with auspicious meaning and special meaning related to religion, power and etiquette.

Various China's ancient accessories

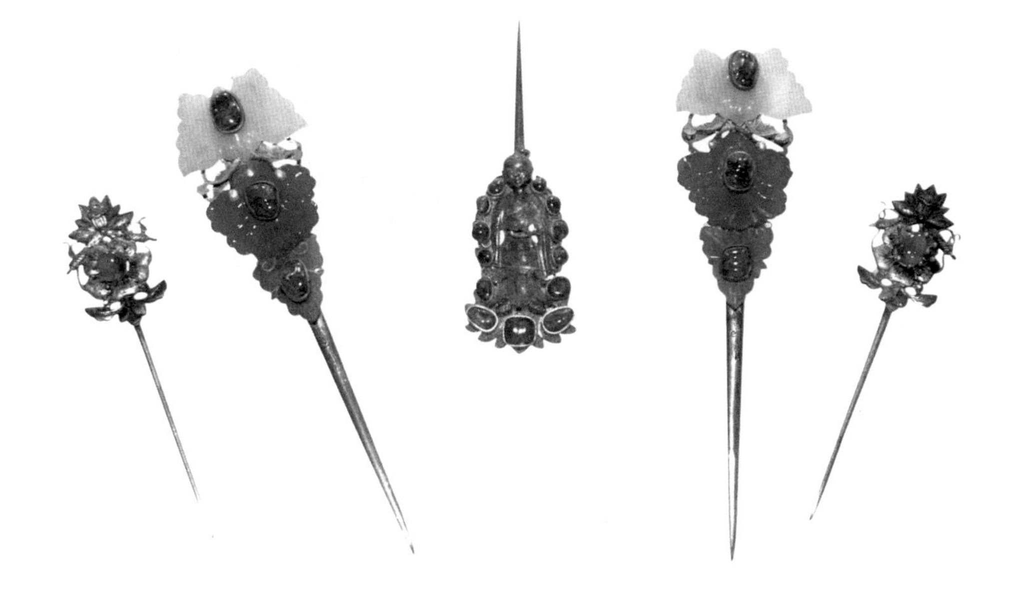

然历经千年岁月，仍风采依旧，体现了与中国传统工艺一脉相承的品质。这些佩饰不仅式样精美，而且具有深厚的文化内涵，很多饰品以动植物、神话传说、历史故事等作为题材，寓意吉祥，记录了中国悠久的历史文化和多姿多彩的民俗风情，融历史、文学、民俗、美术于一体，令人爱不释手，百看不厌。

本书以深入浅出的文字和珍贵精美的图片，向读者展示了中国古代的各种吉祥佩件和精美首饰，使读者领略到古代佩饰的艺术之美，同时更多地了解中国的传统文化与民俗民风。

have been retaining their original exquisite workmanship withstanding the test of time. Their designs deriving from plants, animals, myths, legends, historical stories and auspicious meanings presented the profound cultural connotation. They constituted the mixture of China's history, literature, folk custom and art.

You will find China's ancient accessories with auspicious meanings and fine jewelry from this book. The beautiful pictures make complicated subjects understandable to you so that you will learn more about Chinese traditional culture, folklore and customs when savoring the artistic beauty of China's ancient accessories.

目 录 Contents

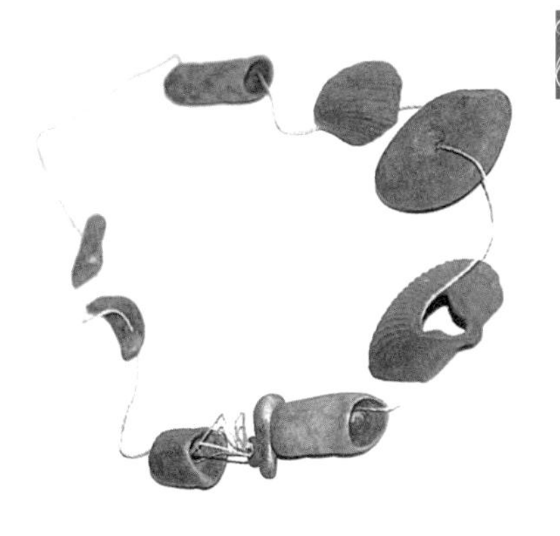

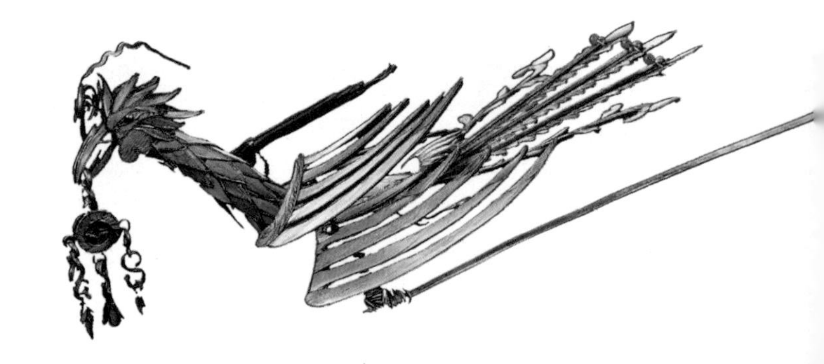

佩饰概说
Introduction of Accessories

佩饰是中国古代服饰的重要组成部分，往往能起到画龙点睛的作用。佩饰是人类审美意识的反映，也体现了人们祈求平安吉祥的美好愿望。

Accessories added the crowning touch to ancient Chinese costumes. As an important part of ancient Chinese costumes, accessories reflected people's common desire for peace, security and propitiousness in addition to being used as ornaments.

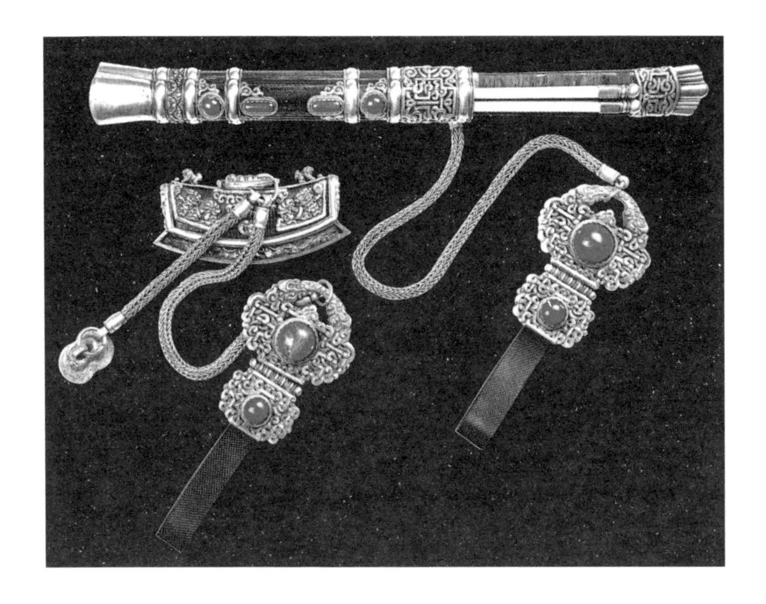

> 佩饰的缘起和发展

追本溯源，佩饰要比衣服出现得更早。早在旧石器时期，古人就将许多小物件佩戴在身上，其材料主要为石英石、砾石、石墨、玛瑙、黑曜石等，还有兽牙和蚌壳等其他材质。当时佩戴的这些小物件有些是作为财富的象征，还有的是一些日常使用的器具和工具，这就是最原始的佩饰。

在距今约1.8万年前的北京山顶洞人的遗址中，就曾发现过一些装饰品，有穿孔的兽牙、海蚶壳、小石珠、小石坠、鲩鱼眼上骨和刻有条纹的骨管等。其中穿孔兽牙最多，有125枚，包括狐狸的上、下犬齿29枚，鹿的上、下犬齿和门齿29枚，野狸的上、下犬齿17枚，鼬的犬齿2枚，虎的门齿1枚等。这些兽

> Origin and Evolution of Accessories

When tracing the origin of accessories, we could notice that the history of wearing accessories was earlier than that of wearing costumes. As early as the Paleolithic Period, primitive people started to wear some small objects made of quartz stone, gravel, graphite, agate, obsidian, animal teeth, clam shells and other materials on their bodies to show off the wealth. Other daily used items and tools became the main ornaments in the primitive society.

Such accessories as bored animal teeth, sea clam shells, stone beads and grass carp fish bone with carved lines have been found in the ruins of Upper Cave Man, about 18,000 years ago. The number of bored animal teeth reached as many as 125 including 29 fox canines, 29 deer laniarii and incisor teeth, 17 wild

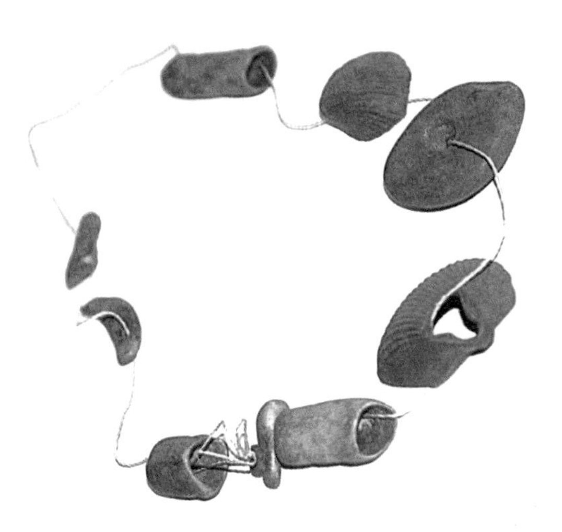

- **北京山顶洞人遗址出土的装饰品（旧石器时代）**

  Unearthed Ornaments of Paleolithic Period from the Ruins of Upper Cave Men (Palaeolithic Age)

牙的根部均钻有一个孔，有一些甚至用赤铁矿粉染成了红色。山顶洞人用绳线将这些小装饰品穿起来，佩戴在脖颈、手腕，或悬挂在衣服上，用以美化外形。可见当时的人类已经有了审美情趣。

新石器时代，人类社会正处于文明起源阶段。“文明”的意义包括农业发展、手工业兴旺、民族形成、城镇勃兴、城乡分离、贫富分化、阶级对立、金属出现、文字发明、文化发达、商品交换出现、国家机构成立等基本内容。当时以农

raccoon canines, two weasel canines and one tiger's incisor. Upper Cave Men strung these teeth which were drilled on the root and dyed red with ferrous powder. They wore them around the neck and wrist or hung on their clothes as ornaments.

Neolithic Age, starting point of civilized society, witnessed unprecedented development in the agriculture, handicraft, culture and commodity exchange. Other events also happened at that time. They were the formation of nations, urban and rural division, polarization between the rich

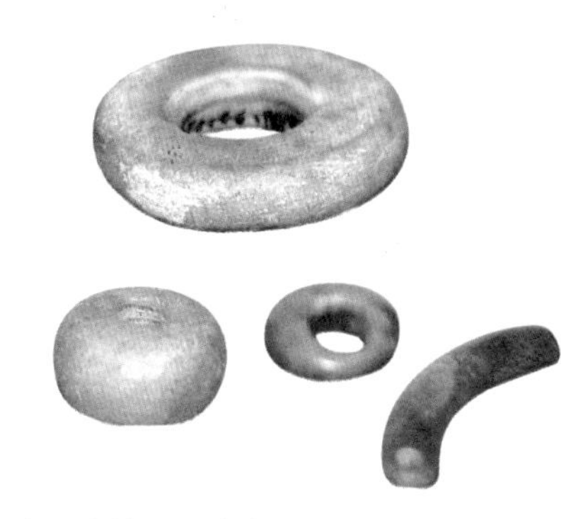

- **浙江余姚河姆渡遗址出土的玉器（新石器时代）**

  余姚河姆渡遗址出土的玉器中有璜、玦、管、珠等多种形制。它们是中国迄今发现的年代较早的玉饰件，多为小件佩饰，且大多光素无纹。

  Unearthed Neolithic Jade in the Ruins of Hemudu, Yuyao, Zhejiang Province (Neolithic Age)

  Unearthed accessories in the ruins of Hemudu, Yuyao were mostly patternless and small-sized jade ornaments such as *Huang*, *Jue*, pipes and beads, which have been one of the earliest types of jade ornaments.

• 山东大汶口文化遗址出土的挂件（新石器时代）

大汶口文化遗址出土的饰品大体可分为头饰（笄）、颈饰（管）和手饰（臂环、手镯、指环）。饰品材质多为石头，其他还有陶、骨、牙、角等。

Unearthed Neolithic Stone Pendants in the Ruins of Dawenkou Culture, Shandong Province (Neolithic Age)

These unearthed accessories included hairpins, necklaces, arm ornaments, bracelets and rings. They were made of stones, pottery, bones, animal teeth, horns and so forth.

耕经济发达为特征的中原地区，较早地进入了原始氏族公社向奴隶制国家过渡的阶段。

这一时期，人们的佩饰更加丰富，形式已不限于项链、腰饰等，还出现了笄、梳篦（梳理头发的用具，也可插在头发上作装饰）、

and the poor, class antagonism, invention of pictograph and the establishment of national institutions. Characterized by the development of agricultural economy in Central Plains region, China stepped into the transitional stage from the primitive clan to slavery system.

During that period, accessories

指环、玉玦、手链等。佩饰的材质也相当丰富，仅出土的梳篦的材质就有骨、石、玉、牙等。另外还出现了一种极具特色的佩饰，被称为“带钩”。带钩就是腰带的挂钩，最初多用玉制成，发展到春秋战国时期最为盛行，材料也更加丰富。

商代（公元前1600—公元前1046）和西周时期（公元前1046—公元前771），佩件以玉佩为主，

became more diverse. They were not limited to the necklace and waist ornaments. Rings, *Jue*, bracelets and combs (either used to comb hair or as hairpins for decoration) appeared. The unearthed combs alone were made of different materials such as jade, animal teeth and bones. A very unique accessory known as "buckle" was used for linking to the belt. Initially it was commonly made of jade. When it prevailed in the Spring and Autumn Period, "buckle" was made of other materials.

Jade ornaments have been made preciously and beautifully since wearing jade ornaments prevailed in the Shang Dynasty (1600 B.C.-1046 B.C.) and the Western Zhou Dynasty (1046 B.C.-771 B.C.). In the Shang Dynasty, jade ornaments were full of aesthetic taste with different decorative patterns

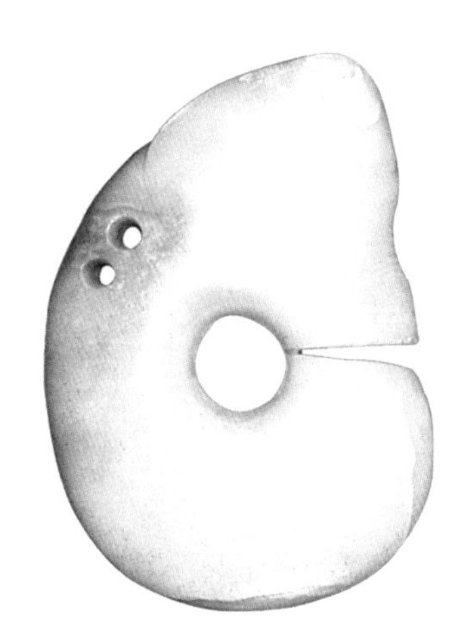

- **内蒙古红山文化遗址出土的“玉猪龙”玦（新石器时代）**

此玉玦的整体形状像字母“C”，中部有一个大圆孔；上部雕成一个与身体不成比例的肥大兽头，嘴部圆长而齐唇，似猪嘴；兽身如圆柱，向前蜷曲，无四肢，尾端与嘴部相接；颈后有两个小穿孔，由两端对钻而成，可穿系绳。所雕之兽，有学者认为是龙，有学者认为是猪，故名“玉猪龙”。

Unearthed Neolithic Jade *Jue* with Mixed Pattern of Pig and Dragon, in the Ruins of Hongshan Culture, Inner Mongolia (Neolithic Age)

This jade *Jue* was in the shape of letter "C" with an engraved beast head out of proportion to the size of its body. It had a pig-like snout in round and long shape with straight lips. Its crouched body was in the shape of cylinder without arm and leg. Its mouth and tail were placed end to end. There were two small drilled holes at the back of the beast for tying ropes. Some scholars believe that it was the prototype of the dragon, but others believe that it was a design of the pig, hence its name.

当时的贵族无论男女都流行玩玉和佩玉，玉饰异常精美珍贵。如商代出现了装饰有人纹、鸟纹、鱼纹、兽纹的佩璜，这是具有审美性质的赏玩性和装饰性的佩玉。这一时期的首饰主要是笄，此外还增加了臂饰，即手臂上的饰品，称为“玉瑗（yuàn）”。在河南安阳商代古墓中就发现有各式的玉瑗：有的形状长而宽，中部有凸棱；有的呈凹弧

such as birds, fish, beasts and human faces. Based on various unearthed arm ornaments in the Shang tomb of Anyang, Henan Province, in addition to hairpins, it was obvious that ornaments on arms were also popular at that time. They were varied in shapes. Some of them were long and wide with convex edge in the middle. Others were in the shape of concave arc with convexity in the middle and bulge on both sides. And there are others like

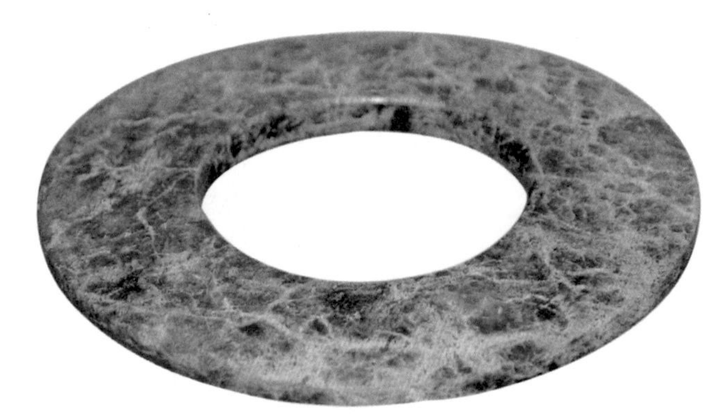

• 浙江良渚文化遗址出土的玉瑗（新石器时代）
Unearthed Jade *Yuan* in the Ruins of Liangzhu Culture, Zhejiang Province (Neolithic Age)

• 浙江良渚文化遗址出土的玉琮式镯（新石器时代）
Unearthed Neolithic Jade *Cong* (Bracelet) in the Ruins of Liangzhu Culture, Zhejiang Province (Neolithic Age)

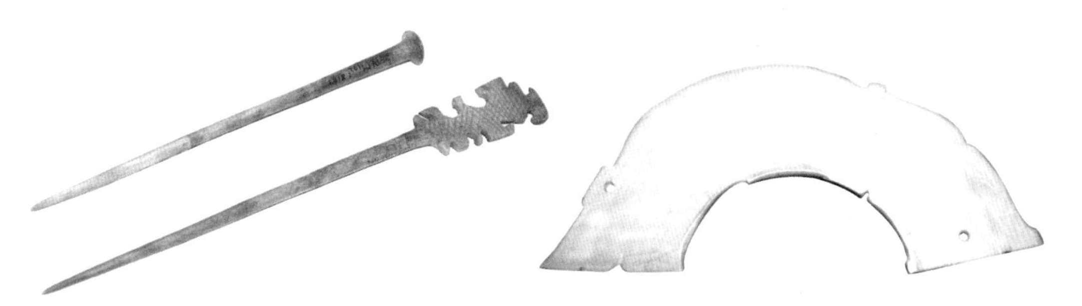

• **骨笄（商）**

商周时期，笄的加工越来越精细，笄首的装饰性也越来越强。笄早已不仅仅是绾发的工具，更成为头部最主要的装饰品和身份地位的象征了。

Bone Hairpin (Shang Dynasty)

The making process of bone hairpins in the Shang and Zhou dynasties became more elaborate. Used as an ornament, the bone hairpin was not only a tool for making hairdo but also an important ornament to show off the wearer's social status.

• **玉璜（商）**

商代出土的玉璜大多类似三分之一的玉璧。

Jade Huang (a Kind of Arch-shaped Jade) (Shang Dynasty)

The size of unearthed jade *Huang* (Shang Dynasty) was only one third of *Bi*, a round flat piece of jade with a hole in its center.

形，中间凹、两边凸；有的内缘凸起，如同碗托。

西周是中国奴隶制王朝的鼎盛时期，大力推行以王权为中心的礼制，在国家礼仪活动中大量使用礼器，尤其是玉礼器，部分玉礼器后来又发展成为玉佩饰。这一时期出现了许多玉佩饰，并首次出现玉组佩。此时的佩饰造型简朴，但纹饰则向繁复和图案化方面发展，常见的有凤鸟纹、夔纹、龙纹、兔纹、鹿纹、象纹等。

bowl holders, bulging on the inner rim.

Carrying out the kingship as the center of ruling system, China's slavery reached the heyday in the Western Zhou Dynasty. Jade sacrificial vessels were widely used in the ceremonial activities. Later on some of them developed into jade ornaments. Strings of jade appeared at that time. Although many jade accessories were simple in shape, patterns tended to be complex. Such patterns as phoenixes, double lines, dragons, rabbits and deer were commonly used.

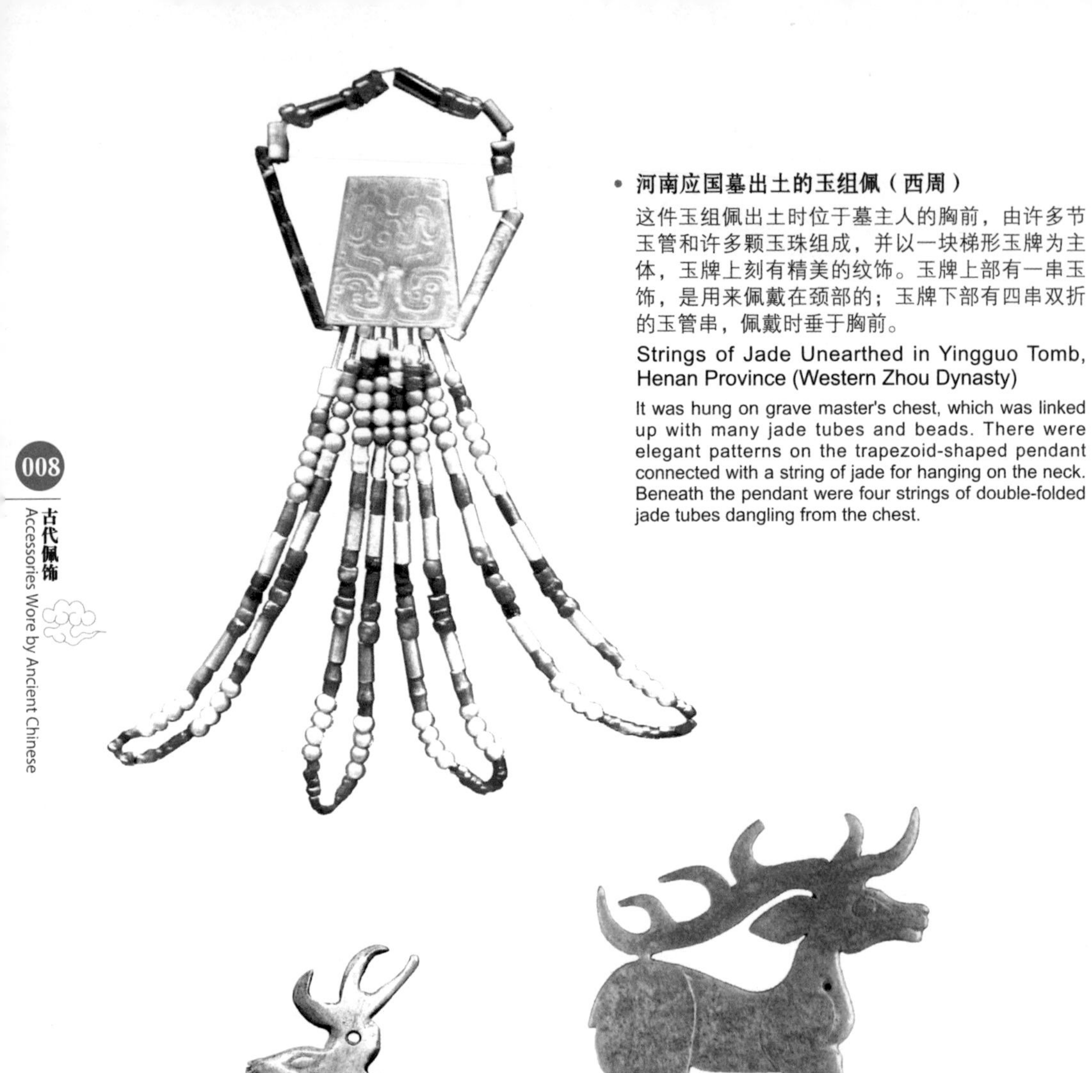

- **河南应国墓出土的玉组佩（西周）**

这件玉组佩出土时位于墓主人的胸前，由许多节玉管和许多颗玉珠组成，并以一块梯形玉牌为主体，玉牌上刻有精美的纹饰。玉牌上部有一串玉饰，是用来佩戴在颈部的；玉牌下部有四串双折的玉管串，佩戴时垂于胸前。

Strings of Jade Unearthed in Yingguo Tomb, Henan Province (Western Zhou Dynasty)

It was hung on grave master's chest, which was linked up with many jade tubes and beads. There were elegant patterns on the trapezoid-shaped pendant connected with a string of jade for hanging on the neck. Beneath the pendant were four strings of double-folded jade tubes dangling from the chest.

- **玉鹿佩（西周）**

西周玉鹿佩多取鹿远眺之状，剪影式造型，简洁生动，极其传神。

Jade Deer-shaped Pendant (Western Zhou Dynasty)

The pattern which was commonly used in the Western Zhou Dynasty depicted a deer in the position of crouching looking far into the distance. It was in silhouette style, simple but vivid.

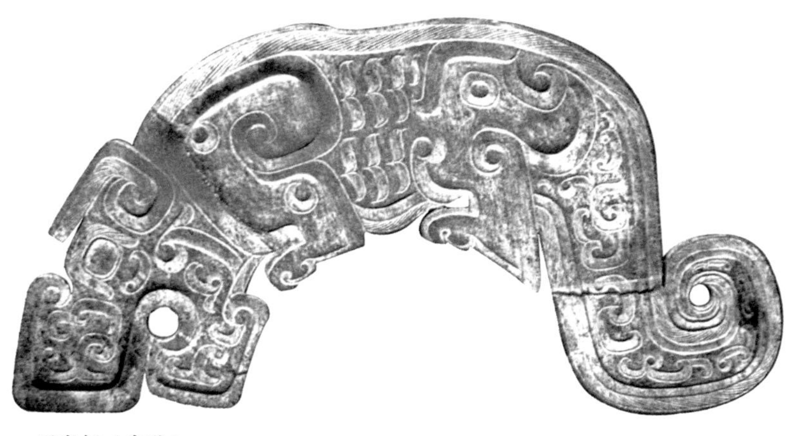

• 玉虎佩（春秋）
Jade Tiger-shaped Pendants(Spring and Autumn Period)

春秋战国时期（前770—前221）是由奴隶制向封建制转化的变革时期。各诸侯王国形成了地域文化圈，因此使各地的佩饰风格迥异，形式大增，各种技法并用。

这一时期，佩饰仍以玉质为主，但形式更加华美，做工更加考究。佩饰普遍带有象征意义，后人可以从出土的饰品中窥探出该饰品主人的身份和社会地位。从春秋战国时期开始，中国出现了开采金、银等金属的矿产业，因此在战国的古墓中，曾发现了大量金、银饰品。

The Spring and Autumn and Warring States Periods (770 B.C.-221 B.C.) witnessed the transitional stage between slavery and feudalism. With the boom in regional culture, every kingdom exerted a great effect on diversified styles of accessories resulting in the integration of different art techniques.

Because of the exquisite craftsmanship and patterns with symbolic meaning, jade accessories dominated any other accessory during that period. The wearer's social status could be identified based on the unearthed accessories. Gold and silver mining industry emerged in the Spring and Autumn Period so a large

这一时期还十分盛行戴玉玦。这种饰品有圆形缺口、素面无纹的，也有雕琢纹饰的，还有呈柱状、有缺口的。带钩亦极其盛行，所用材料有金、银、玉、青铜等，制作十分考究。其装饰工艺除雕镂花纹外，还有在青铜上镶嵌绿松石，或在铜、银上鎏金等。春秋战国时期的佩璜纹饰日趋繁复，题材多为龙、凤、蟠螭、云纹等，周身施饰。

number of gold and silver accessories were found in the tombs of the Warring States Period.

Jade *Jue* also prevailed in that period of time. Some were in penannular shape with or without patterns. Others were in columnar shape with an opening. The buckle accessory made of gold, silver, jade, bronze and other materials was also popular. In addition to hollow-out carving patterns, some bronze buckles were inlaid with turquoise. Others were gilt on copper and silver. The techniques were extremely exquisite. Diverse patterns on the jade *Huang* of the Spring and Autumn Period tended to be more complex, such as dragon, phoenix *Panchi* and cloud patterns.

河南信阳出土的彩绘俑复原图（战国）

图中二人身穿交领右衽直裾袍，袖口呈弧状，饰菱纹。腰上佩挂由穿珠、玉璜、玉璧、彩结、彩环等组成的组佩。

Restoration Painting of Colored Drawing Pottery Unearthed in Xinyang, Henan Province (Warring State Period)

Two figures in the drawing wore crossed collar and right-buttoned garment with loose arced sleeves. It was decorated with diamond-shaped edge, beads, colored ribbons and groups of jade ornaments.

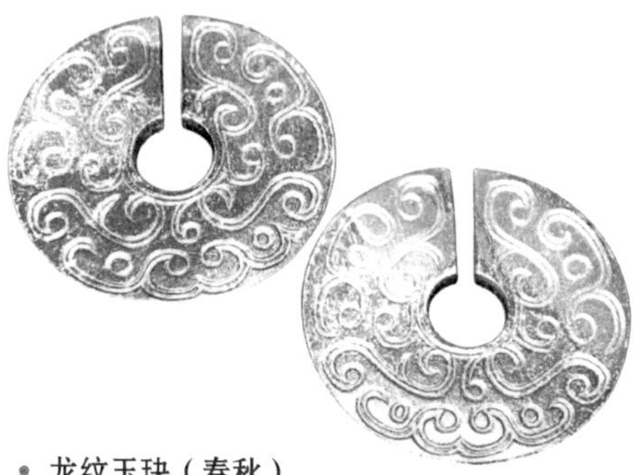

• **龙纹玉玦（春秋）**

春秋时期的玉玦，纹饰主要为勾连纹和以弧线为主的简化龙纹。用来表现龙身的阴刻弧线，又长又匀，与其边缘呈同心圆状。

Jade *Jue* with Dragon Patterns (Spring and Autumn Period)

Such simplified dragon patterns as arc and hook shaped design were commonly used on the jade *Jue* during the Spring and Autumn Period. This design depicted the long and even intaglio curve on the dragon body.

• **双龙首玉璜（春秋）**

璜的外沿有数个凹形缺口。纹饰一般为龙纹或兽形纹。

Jade *Huang* with the Pattern of Two Dragon Heads (Spring and Autumn Period)

It was decorated with the dragon or zoomorphic pattern. Several concaves were presented on outer edges.

• **绳纹玉环（战国）**

绳纹又叫“绞丝纹”，指两股绳索绞扭的形状。每一股绳索由二条、三条、四条甚至九条单线绞扭而成。

Jade Ring with Rope Pattern (Warring States Period)

It referred to the pattern of two twisted ropes. Each rope was made of two, three, four or nine strands of strings.

• **鎏金银带钩（战国）**

战国中后期，带钩的制作与使用进入鼎盛期。新兴的琵琶形带钩空前盛行，钩身上的浮雕式兽面纹，立体感甚强。

Gilt Silver Belt Hook (Warring States Period)

Hook making peaked in the mid-late Warring States Period. Newly emerged lute-shaped hook reached unprecedented prevalence. Animal face patterns were embossed, full of three-dimensional sense.

• 三龙璧形玉佩（战国）

此玉佩是玉雕镂空技艺中的珍品，扁圆形，两面纹饰相同，中央为小环，璧外缘分布三条相同卷曲扭动的龙，首尾相连，呈“弓”形，是由“S”形龙纹演化而来的。

Jade Round Flat Pendant with Three Dragons (Warring States Period)

It was a treasure of hollow-out carving. Both sides had the same patterns. It was a flat piece of jade with a hole in its center. Three wiggly dragons around the outer hole were end to end presenting the shape of "S", which was one of the innovations of dragon patterns.

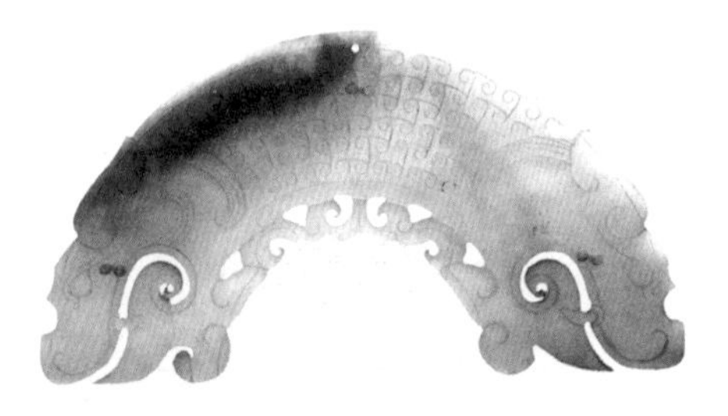

• 双龙首玉璜（战国）

此玉璜的基本形制是圆璧的二分之一，璜体较宽，在璜的两端雕刻两个侧面龙头，弯月形的璜体上饰谷纹或云纹。在璜体下内沿，有镂空装饰。

Jade *Huang* with a Dragon Head on Each End (Warring States Period)

It was about the half size of the round-shaped jade *Bi*. A dragon head was carved on each end of the crescent-shaped jade. The whole piece was decorated with cloud pattern or cereal pattern, the hollow-out carving pattern beneath.

• 虎食人纹玉佩（战国）

此玉佩通长6.2厘米，宽3.8厘米，厚0.4厘米，两面雕刻，线条细如毫发。玉佩的中心采用虎食人纹，即为一猛虎作食人状，虎爪分别按住人的手脚，张口咬住人的腰腹部，人奋臂伸足作挣扎状。环的两侧各饰一个扬臂翘足的舞人形象，舞人脚下又各有一个蛇形物。

Jade Pendant with Tiger Eating Man Pattern (Warring States Period)

It was 6.2 centimeters in length, 3.8 centimeters in width and 0.4 centimeter in thickness. Engraving lines were as fine as silks on both sides of the pendant. On the center showed the pattern of a ferocious tiger biting a man. The man struggled mightily when his hands and feet were pressed by the tiger and his waist and abdomen were bitten. There was a dancer on each side with a snake beneath his feet.

• **龙形玉佩（战国）**

此玉佩采用了阴刻、单面雕、双面雕、镂空雕等多种技艺。

Entwining-dragon-shaped Jade Pendant (Warring States Period)

It was carved in a variety of techniques, such as intaglio, one-sided engraving, double-sized engraving and hollow-out carving.

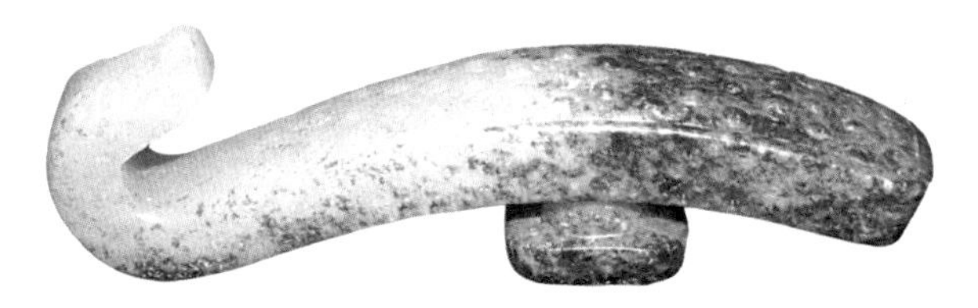

• **白玉带钩（战国）**

战国时期的玉带钩大的长22—24厘米，用来束衣；小的长4—6厘米，卡在革带上供挂物之用。

White Jade Belt Hook (Warring States Period)

It was used to hook the belt for hanging things with the longest length 22 centimeters to 24 centimeters, the shortest one 4 centimeters to 6 centimeters.

秦朝（公元前221—公元前206）统一中国后，清除异制，建立了自己的佩饰文化，但佩玉制度继承前代。由于秦朝统治的时间很短，故流传下来的佩饰极少。

汉代（公元前206—220）的佩饰是在长达四百多年的大一统封建

With the unification of China in the Qin Dynasty (221 B.C.-206 B.C.), the unified accessories system was established after the removal of inconsistent system; however, the jade related system retained. Because the Qin Dynasty existed for a very short period, very few accessories were handed down.

王朝背景下发展起来的，因此出现了佩饰发展的高峰。此外，汉代始通西域，开通了丝绸之路，新疆出产的和田玉得以大量进入中原地区。白玉的使用开始兴

The accessories made headway in the reign of the Han Dynasty (206 B.C.-220 A.D.) which lasted more than 400 years. Accessories development reached a peak in the Great Unification of the feudal dynasties. In addition, the Western Han Dynasty began to establish the Silk Road leading to the Western Regions. A great deal of Hetian jade (a kind of nephrite, is one of China's five most famous jades) which was produced in Xinjiang poured into the Central Plains. Thereafter white jade began to flourish and it was deeply loved by Chinese people at that time.

Accessories of the Qin and Han dynasties followed the style of previous

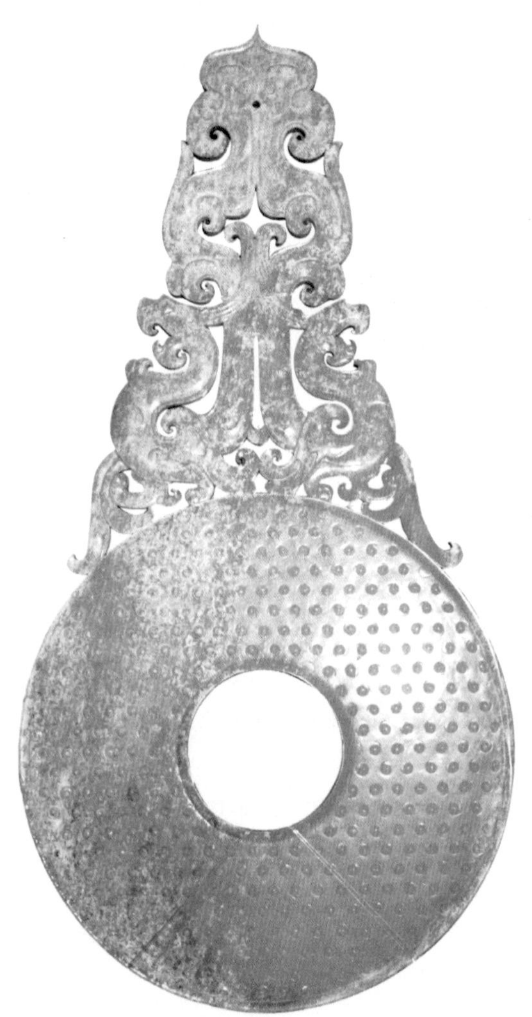

- **透雕双龙出廓玉璧（西汉）**

  出廓璧是在圆廓之外多出一部分镂雕装饰的玉璧，最早产生于战国时期。此玉璧的出廓部分仅在玉璧的一侧，且伸出较长，镂雕勾连纹和龙纹。

  Jade Ornament *Bi* with Hollow-out Carved Twin Dragons (Western Han Dynasty)

  It appeared in the Warring States Period. There were hollow-out carved dragons stretching out from one side of the round flat piece of jade.

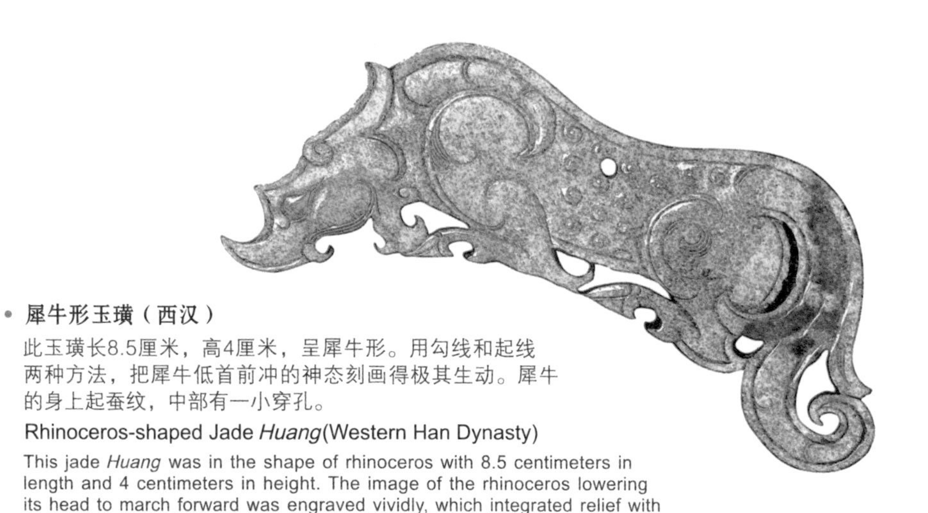

- **犀牛形玉璜（西汉）**

  此玉璜长8.5厘米，高4厘米，呈犀牛形。用勾线和起线两种方法，把犀牛低首前冲的神态刻画得极其生动。犀牛的身上起蚕纹，中部有一小穿孔。

  Rhinoceros-shaped Jade *Huang*(Western Han Dynasty)

  This jade *Huang* was in the shape of rhinoceros with 8.5 centimeters in length and 4 centimeters in height. The image of the rhinoceros lowering its head to march forward was engraved vividly, which integrated relief with intaglio for detailed presentation. There was a hole in the middle.

- **龙形玉环（西汉）**

  Dragon-shaped Jade Ring (Western Han Dynasty)

- **六棱橄榄形红玛瑙珠链（西汉）**

  红玛瑙是以颜色命名的玛瑙品种之一，古称“赤玉”。

  Hexagonal Red Agates Chain (Western Han Dynasty)

  The red agate was referred to as "red jade" because of its color.

盛，成为当时中国人最喜爱、最推崇的玉料。

秦汉时期，佩饰的样式虽多承袭前代，但制作材料较前代更加丰富，工艺和形制也更为精巧，造型、质料、色彩上的差别亦愈加明显。贵族男子多随身佩戴印章、

dynasties; however, materials, colors and shapes became more diverse. Workmanship tended to be more complicated and ingenious. Aristocratic men started to wear seals, swords and pouches which were attached to the belt. Strings of ornaments hanging on the ribbons became popular at that time. Men

- **龙形玉环（汉）**

此玉环直径 7.9厘米，厚0.3厘米，为扁平体，透雕，两面纹饰相同。夔龙的身体弯曲，头尾相接呈环形。龙首较长，尖耳上翘，圆眼，巨口大张，上唇长于下唇，下唇外卷，并以利齿衔于尾部。龙身阴刻卷云纹，身外还有透雕的弯钩状饰物，似鬣毛。

Dragon-shaped Jade Ring (Han Dynasty)

It was a flat piece of jade using openwork carving technique with the same pattern on both sides. It had a diameter of 7.9 centimeters and a thickness of 0.3 centimeter. The dragon bending the body end to end formed a circular shape. It had a longer head with sharp ears upturned, round eyes and a big open mouth. Its upper lip was longer than turned outward lower lip. The sharp teeth held the tail. Intaglio cloud patterns were engraved on the body of the dragon with hook-shaped openwork carving ornaments, similar to manes.

- **双龙衔环玉璧（汉）**

玉璧是一种圆板形、片状、中部有孔的玉器。其用途很广，可做祭器、礼器、佩饰、衡器。此玉璧高30 厘米，璧径24.4厘米，厚1.1厘米。廓较宽较高，廓内雕有谷纹，廓外两侧雕刻一对镂空的小龙。

Jade *Bi* with Two Dragons Holds a Ring in Mouth (Han Dynasty)

Jade *Bi* is a circular flat piece of jade with a hole in its center, which was widely used as sacrificial vessels in the ceremony of offering sacrifices, ornaments and scales. The jade was 30 centimeters in height, 24.4 centimeters in diameter and 1.1 centimeters in thickness. There was a small hollow-out carved dragon on each side. Cereal patterns were engraved on the main body.

• 银印（汉）

印章是古代用于辨识身份的工具，也是地位和权力的象征，一般会穿上绶带以方便佩戴。

Silver Seal (Han Dynasty)

It was a tool used in ancient times to identify people's social status. It generally went with the ribbon.

鞶（pán）囊（佩在腰带旁的小口袋）、刀、剑等物，并且大都佩挂于腰部的革带上，组绶串饰也开始流行。首饰方面，汉代男子一般只用笄，女子除笄以外还用钗、簪、步摇和耳珰（dāng）。汉钗的形状比较简单，是将一根金属丝弯曲为两股而成。耳珰多作腰鼓形，一端较粗，常凸起呈半球状。戴的时候以细端塞入耳垂的穿孔中，粗端留在耳垂前部。还有一种在其中心钻孔、穿线、系坠的耳饰，称为“珥”。

in the Han Dynasty preferred wearing large-sized hairpins while women wore hair clasps, hairpins with hanging beads and eardrops besides large-sized hairpins. The shape of the hairpin in the Han Dynasty was relatively simple and it was made of a wire bent into two strands. Eardrops looked like a drum with a hemispherical shape on one thick end. When wearing them, wearers inserted the thin end into ear lobe leaving the thick end on front. Another style was to thread through the ear lobe attached to ear pendants.

丝绸之路

在东西方经济、文化交流史上，有一条举世闻名的丝绸之路。丝绸之路开通于汉代。当时，汉朝军队在西域打败匈奴，控制了张掖、酒泉等关口，并派出了丝绸商队到达安息（伊朗高原古代国家）。商队促进了中原与西域之间的交流。后经商人们的西进或东行，丝绸之路延伸到帕米尔高原以西的中亚、印度、西亚乃至欧洲。丝绸之路开通后，珠宝、植物、金银器等进入中国，而中国的丝绸、瓷器、茶叶、铁器、漆器、指南针、火药、造纸术、印刷术等也通过丝绸之路被带到中亚等地。

丝绸之路不仅是沟通中国和中亚、西亚及欧洲各国的交通商道，还是各国经济、文化、科学技术相互往来的桥梁。东西方物质文明和精神文明相互交流，促进了全人类社会的不断发展和科学技术的不断进步。

• 绸缎
Silk Fabrics

The Silk Road

China opened a well-known Silk Road in the Han Dynasty which made contribution to cultural exchange and economy between the East and the West. After the Hun army was defeated, such strategic passes as Zhangye and Jiuquan were controlled by army of the Han Dynasty. And then the government of the Han Dynasty trade Caravans bringing silks arrived in Anxi (an ancient country on Iranian plateau), which promoted business relationship between the Central Plains and the Western countries. Thanks to trading between western and eastern merchants, the Silk Road extended to the west of Pamirs, Central Asia, Western Asia, India and Europe. Jewelry, plants, gold and silver vessels poured in China via the Silk Road while China's papermaking and printing techniques and products such as silks, porcelains, tea, lacquers, compasses and gun powder were transported to Central Asia and other places.

The Silk Road acted as a bridge linking trading, culture, science and technology between China and other areas such as Central Asia, Western Asia and European countries. It helped to promote not only the mutual exchanges in material and spiritual civilization but also the continuous development of science and technology.

• 云龙纹漆屏风（西汉）

Lacquer Screen with Cloud and Dragon Pattern (Western Han Dynasty)

汉武帝与玉搔头

从汉代开始，玉簪有了一个文雅的名称，叫作“玉搔头”。关于这个名称，还有一个有趣的典故。汉武帝有一位宠爱的妃子——李夫人，原本是一个舞伎，出身低微。李夫人的哥哥李延年能歌善舞，有一次在汉武帝面前起舞唱道：“北方有佳人，绝世而独立。一顾倾人城，再顾倾人国。宁不知倾城与倾国，佳人难再得！”汉武帝非常好奇，于是召见这位佳人，并将她纳为妃子。据史料记载，汉武帝有一天到李夫人宫中，觉得头皮发痒，便拿起李夫人头上的玉簪搔头。于是此后，宫中的嫔妃都将玉簪称作“玉搔头”。汉代以前的簪多为骨制，但从此以后，开始盛行用玉作簪，致使玉石身价百倍。

- **汉武帝蜡像**

  汉武帝是汉朝的第五位天子，十六岁登基，在位五十四年。他开拓了汉朝最大的版图，功业辉煌。

  Wax Figure of Emperor Wudi of the Han Dynasty

  Emperor Wudi was the fifth emperor of the Han Dynasty. He throned at the age of 16 reigning from 140 B.C. to 87 B.C. He made brilliant achievement by expanding the farthest territory of the Han Dynasty.

Emperor Wudi of the Han Dynasty and Jade Hairpin

The story goes that in the Han Dynasty the jade hairpin nicknamed "scratching the head". Emperor Wudi of the Han Dynasty had a beloved concubine, Ms. Li. She was born in humble. Her brother, Li Yannian, was very good at singing and dancing. Once he sang for the Emperor Wudi of the Han Dynasty, "There was a beauty in the north of China. She was so beautiful that put all beauties in the shade. If you haven't met her before, you would never ever get the real beauty." The song aroused his great curiosity. She was summoned to the presence of the Emperor Wudi of the Han Dynasty and was given the title of concubine. According to historical records, on arrival of her palace Emperor Wudi of the Han Dynasty felt itchy so he picked up her jade hairpin to scratch his head. Thereafter, concubines named it "scratching the head". Since then jade hairpins were popular resulting in the rocketing price Besides, most of hairpins were made of bones before the Han Dynasty.

在魏晋南北朝近四百年的时间里，政局动荡、战乱不息、经济萧条，佩饰的造型、工艺和艺术风格均沿袭前代，传世之物较少。

The Wei, Jin, Southern and Northern dynasties (220-589) lasted nearly four hundred years. Because of political instability, endless wars and economic depression, the accessories of that period of time followed the styles and techniques of previous dynasties. Not many accessories were passed down.

• **螭纹镙形玉佩（南北朝）**

此玉佩高6.25厘米，宽5.65厘米，厚0.55厘米，造型仍为汉代形制，但螭身及腿部有表示毛的密排短线，螭身上还有小圆圈。

Jade Pendant with Dragon Pattern (Northern and Southern Dynasties)

It was 6.25 centimeters in height, 5.65 centimeters in width and 0.55 centimeter in thickness. Its shape remained same as the one in the Han Dynasty but the difference lied in the pattern. The short and thick lines were on the dragon's legs and small circles on the dragon's body.

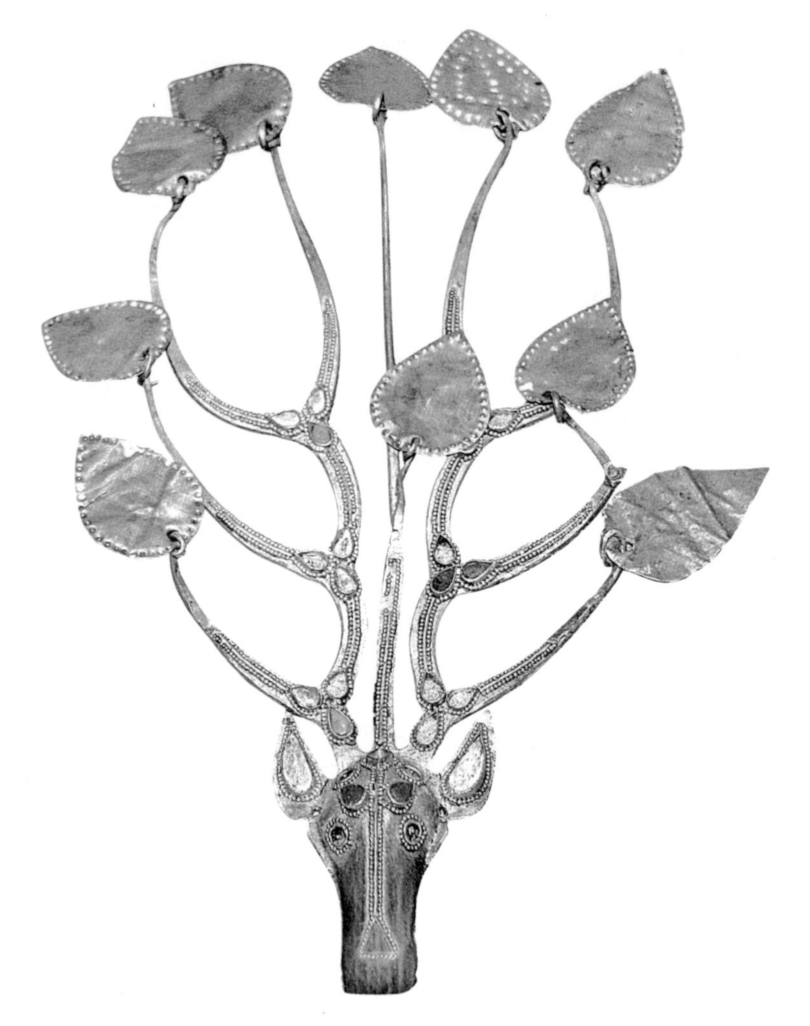

- **马头鹿角形金饰件（南北朝）**

此饰件轮廓清楚，工艺十分精湛。马头部狭长，鼻尖镶白色料石，鼻梁上端用狭薄条金片圈成一个菱形装饰，桃形眉饰镶有料石，眉梢上端另加一对圆圈纹。此饰件有三个形似树枝的鹿角，中间一个无分枝，枝干下部饰一圈鱼子纹，每个鹿角上都悬挂有桃形叶片。

Gold Accessory with Horse Head and Deer Antler Pattern (Northern and Southern Dynasties)

This accessory was a piece of exquisite workmanship. The horse head was in narrow and long shape. The tip of the nose was inlaid with white stones. Narrow and thin gold films circled the top of the bride of the nose in a diamond shape. Peach-like eyebrows were inlaid with stones. A pair of circular patterns were used on the tip of the brows. Three antlers erected upwards with a stem-like antler in the middle. Peach-like leaves were hung on the antlers circled by fish-pattern decoration.

- **螭纹玉璧（南北朝）**

蟠螭是龙的一种，简称“螭”，生得虎形龙相，具有龙的威武和虎的勇猛。此璧以白玉制成，直径9.5厘米，两面均有纹饰：一面是蒲纹，一面是表现螭在水中隐现出没的双螭纹和云水纹。螭纹的线条十分流畅，但健劲之气不及汉代。

Jade *Bi* with Dragon Patterns (Northern and Southern Dynasties)

Panchi, a kind of dragon, resembled the dragon with tiger's build so it was equal parts of tiger's fierceness and dragon's mightiness. This pendant, 9.5 centimeters in diameter was made of white jade decorated with patterns on both sides. Cattail patterns were engraved on one side. There were cloud patterns and dragon patterns on the other side to present the image of the dragon looming in the water. The engraving ran flexibly but lack of force compared with the engraving in the Han Dynasty.

隋唐时期是中国封建社会的强盛时期，统一的多民族国家进一步发展，社会经济呈现出前所未有的繁荣景象。统治阶级为了追求豪华的生活，大量使用金银制作饰物。隋王朝（581—618）历时较短，迄今为止仅发现一些手镯、项链、戒指等。唐代（618—907）奢靡享乐之风盛行，佩饰品种丰富，尤其是金银饰品，成为展示大唐王朝繁荣昌盛、灿烂夺目的标志之一，也是贵族阶层高贵身份的一种象征。同时，开放的大唐帝国吸收和接受了外来文化，中西方文化交流更加频繁，佩饰的造型和纹饰也随之出现了新的面貌。

The Sui and Tang dynasties (581-907) witnessed the prosperous period of China's feudal society. A unified country led to further development and brought about the unprecedented prosperity in social economy. The ruling class was in pursuit of an extravagant life so a great number of gold and silver ornaments were widely used. Because the reign of the Sui Dynasty (581-618) did not last long, the accessories we have found so far were no more than some bracelets, necklaces and rings. Followed the extravagant tendency, accessories were more diversified. Gold and silver ornaments in particular were used to present the magnificent power of the Tang Dynasty. They were also used

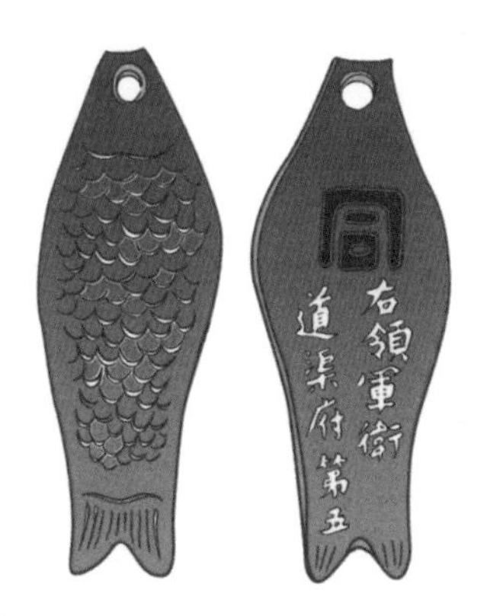

鱼符

鱼符是唐代朝廷颁发给官吏的鱼形符信，鱼符的质料因官阶不同而有所区别。鱼符上面刻有文字，分成两爿，一爿在朝廷，一爿自带。官员迁升或是出入宫廷，以鱼符为凭信。

Fish-shaped Tallies

The imperial court of the Tang Dynasty issued fish-shaped tallies to officials. Material varied with ranks of officials. There were engraved characters on the fish-shape tallies. It fell into two slit bamboos or chopped wood. One half was kept by the court and the other half was kept by the official. Two halves had to be tallied with each other before entering and going out of the imperial court.

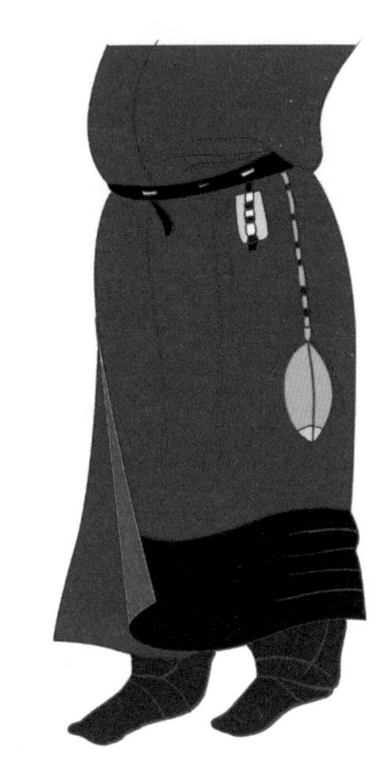

鱼袋佩戴示意图

Sketch Map of Wearing the "Fish Bag"

唐代初年曾流行过一种蹀躞（xiè）带，带上挂有“七事”。此外，朝廷实行“佩鱼”之制，即官员佩戴鱼袋，用以证明官员的身份。唐高宗时，开始赏赐五品以上官员鱼袋。鱼袋饰以金银物品，内装鱼符，以方便查验官员身份。官员出入宫廷时，朝廷要对所有官员进行检查，以防止身份作假。

as showing off the noble status of the aristocracy. The Tang Dynasty was open enough to absorb western culture in the frequent cultural exchange, which help to make the accessories more diverse.

The imperial court started to implement the institution of "wearing the fish bag" (a kind of pocket for holding fish-shaped tallies) in the early years of the Tang Dynasty. It was used to

在唐代，女子用镯子装饰手臂已很普遍，当时人称之为“臂钏”，这与当时吊带衫、无袖衫的盛行有关。当时的臂钏由捶扁的金银条盘绕旋转而成，呈弹簧状，少则三圈，多则五圈、八圈、十几圈不等。此外，唐代女子还盛行在发上插戴梳篦。最初只在髻前单插一梳，后来数量渐渐增加，以两把

identify the officials. Officials above the fifth ranks were rewarded a "fish bag" with fish-shaped tallies decorated with gold and silver in the reign of Emperor Gaozong of the Tang Dynasty. All officials were asked to show their tallies to the guard when going out and in the court.

Because the sleeveless garment prevailed in the Tang Dynasty, armlets

- 《步辇图》中戴臂钏的宫女（唐）
The Court Ladies Wearing Armlets in the Painting of *Emperor Taizong's Sedan* (*Bunian Tu*) (Tang Dynasty)

梳子为一组，上下相对插戴。至晚唐时，贵族女子常常在髻前及其两侧插三组梳篦，梳背的装饰亦日趋富丽。

• **狮纹玉带板（唐）**

唐代玉带板有方形和椭圆形两种，纹饰有人物、花卉、动物等。人物纹多为手持乐器的乐伎，还有朝贡的胡人等。

Jade Belt Plate with Lion Pattern (Tang Dynasty)

It had two types. One was square in shape. The other was oval in shape. It was decorated with different patterns such as figures, flowers and animals. Such figure patterns as musicians holding musical instrument and people paying tribute were commonly used.

were commonly worn by women to decorate arms. Armlets were made of flattened gold and silver bars in spring-like shape varying from three coils, five coils and eight coils to dozens of coils. Wearing combs was popular among women in the Tang Dynasty. A comb was inserted in front of the bun but later the number of comb was increased. Two combs as a pair were inserted from upper and lower directions at the same time. Aristocratic women in the late Tang Dynasty wore three pairs of combs. One pair was placed in the front of the bun. The other two pairs were worn beside each side of the bun. Accordingly, patterns on the back of the comb became increasingly exquisite.

• **花卉纹玉梳背（唐）**

梳背是镶嵌在银梳子或木梳子上的饰件，呈半月形。下部边沿处那一道较薄的边棱，是用于镶嵌的嵌口。一般梳背均两面有浅浮雕的牡丹纹、凤纹和孔雀纹等，刻画细致。

The Back of the Jade Comb with Floral Pattern (Tang Dynasty)

The back of the comb was embedded in silver or wooden comb. It was in half-moon shape with a thin edge for comb inlay. There were peony, phoenix and peacock patterns on both sides. Bas-reliefs were very detailed.

蹀躞七事

“蹀躞七事”是一种简称，指五品以上武官的腰带上佩带的七件物品，包括佩刀、算囊（贮放物品的袋子）、砺石（磨刀石）、契苾真（用于雕刻的楔子）、哕厥（用于解绳结的锥子）、针筒和火石（点火工具）。“蹀躞七事”在唐代极为兴盛，并沿用至北宋年间。

“蹀躞七事”来源于唐代时中亚地区，那里的骑士在出征的时候除了腰间悬挂五件自卫武器与行军必备用具外，还要另外佩戴两件有“护身符”作用的物件，以满足“七事”的数量，象征吉祥。

唐朝开国皇帝李渊任山西河东慰抚大使期间曾与突厥、粟特等族发生过交流和冲突。这些外邦将士骁勇善战，佩带的“七事”也十分引人注目。于是，唐军中很多将士也开始佩带“七事”，后来逐渐成为一种流行的军中装束，称为“蹀躞七事”。

唐朝中后期，统治者规定一般官员不再佩挂“蹀躞七事”。但在民间，许多女子喜欢这种佩饰，只是省去了“七事”，以窄皮条代替，仅存装饰之意，无实际的使用价值。

Seven Parts Attached to the Belt

Military officials above the fifth ranks wore the belt with seven parts attached. "Seven parts" referred to the walking sabre, the bag for storage, the grinding stone, the wedge for carving, the

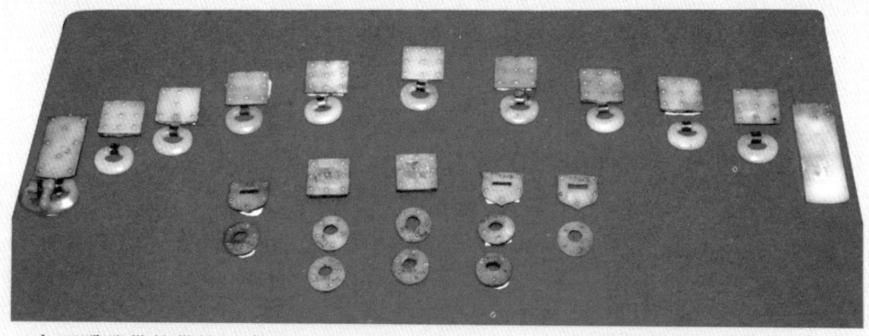

• 九环蹀躞带的带饰（唐）

Ornaments on the Strap (Tang Dynasty)

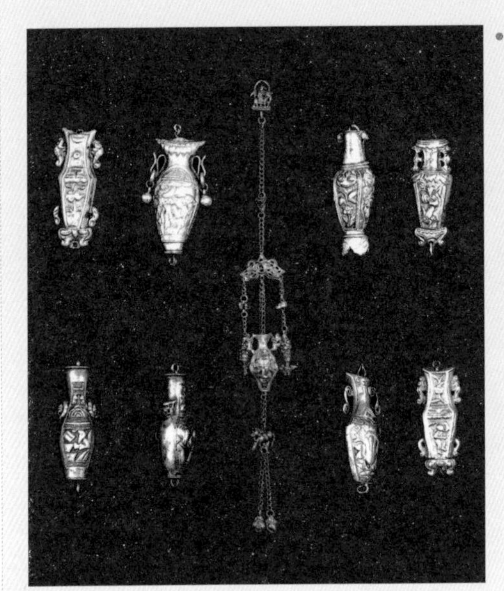

“四季平安”银针筒（清）

针筒即放针的筒，也有一些针筒由竹子制成，用于放纸张或帛书。古代的针筒既是实用品，也是一串美丽的饰品。针筒挂在腰间，走路时小银件相互撞击，会发出声响，十分悦耳。此针筒以花瓶造型制作，“瓶”与“平”同音，寓意平安，錾刻的梅花、牡丹、荷花和菊花则代表四季。

Silver Needle Holder with Engraved Patterns (Qing Dynasty)

Some of needle holders were made of bamboo to hold paper and silk manuscripts. Needle holders in ancient times were both practical and artistic. Hung the needle holder on the waist, it would make musical sound with each footstep. This needle holder was vase-shaped. Based on Chinese pronunciation, "vase" (瓶, *Ping*) is a homonym of "safe and sound" (平, *Ping*). And the engraved plum blossom, peony, lotus and chrysanthemum stood for four seasons respectively.

awl for untying knot, the silver needle holder and the flint. They were very popular in the Tang Dynasty and have been used until the Northern Song Dynasty.

In the Tang Dynasty, "seven parts" were originated from Sute nationality of central Asia where knights wore five self-defense weapons and necessities for marching besides two amulets totaling seven, a lucky number.

Before being the first emperor of the Tang Dynasty, Li Yuan served as an ambassador in Shanxi province. In the conflicts with Sute nationality, Sute knights showed valor and skill on the battlefield and "seven parts" with them were eye-catching. With its popularity among military officials in the Tang Dynasty, "seven parts" gradually became a popular ornament decorated on the military attire.

According to the regulation released by the ruler of the mid-Tang Dynasty, officials were not allowed to wear "seven parts"; however, they were loved by common people. Women replaced "seven parts" with a narrow leather strap for ornament but the original function of "seven parts" no longer existed.

宋代（960—1279）佩饰的材质主要包括玉、金、银、水晶、宝石等，十分丰富。受皇家复古艺术风气的影响，仿古佩饰盛行，追求朴质无华、平淡自然的情趣意味，反对矫揉造作的富丽繁缛之风。此外，宋代佩饰的装饰题材也空前丰富，洋溢着浓郁的生活气息。

宋代官员沿用唐代的“佩鱼”制度，赐予着紫、绯色公服的官吏（高品级官吏）鱼袋。不过此时鱼袋已渐渐失去了实际用途，只用以装饰。鱼袋的形式更为多样，甚至装饰有金银。还有一种“顺袋”，也是人们经常佩带之物，由于其外形被裁制成弧形，状如茄子，因此又被称为“茄袋”。

Materials of accessories in the Song Dynasty (960-1279) were diverse including jade, gold, silver, crystal and gems. Affected by the prevalence of retro artistic style in the royal palace, accessories at that time were in pursuit of primitive, simple and natural style, which was against the extravagant decoration as used to be. In addition, patterns on the accessories became unprecedentedly diverse, full of a strong flavor of daily life.

Following one of the institutions of the Tang Dynasty, officials in the Song Dynasty who were bestowed purple and scarlet official attires (they were high-rank officials) wore "fish bag"; however, "fish bag" at that time has lost its practical use. It was only used for ornament. A variety of shapes appeared and some of "fish bag" were decorated with gold and silver. A kind of bag in arc shape like an eggplant was commonly used, hence its name.

• 双龙纹玉镯（宋）

Jade Bracelet with Twin Dragons Pattern (Song Dynasty)

- **双鹤衔草纹玉佩（宋）**

双鹤衔草纹玉佩是宋代传世品中最常见的佩玉。此玉佩高4.3厘米，宽6.8厘米，以双鹤对飞之形为玉佩。玉鹤颈长而曲，双鹤头相对，喙相接，短翅平展并向前伸，两鹤内侧之翅相接，翅上有细长的平行阴线羽，翅之边缘如锯齿。两鹤的足下饰云纹，云纹如蔓草卷向两侧。双鹤头的后面有一圆环用于系绳。

Jade Pendant with Two Cranes Holding Grass Pattern (Song Dynasty)

It was the most common jade pendant handed down in the Song Dynasty. It showed the design of two cranes flying together with 4.3 centimeters in height and 6.8 centimeters in width. Two cranes, face to face, beak with beak, spread their wings forward with serrated border. Their inner side of wings met together presenting intaglio patterns of paralleled slender features. Their feet met together with cloud patterns beneath. Cloud patterns were just like creeping weeds which spread to both sides. There was an annular hole at the back of the head for string.

- **吹笙乐伎纹玉带板（南宋）**

玉带板即腰带上的玉饰。唐代以来，对官服腰带上的饰物和材质做了规定，以区别官职的高低，其中玉制的带板是最尊贵的。历代的玉带板做工、纹饰都不相同，有鲜明的时代特征。

Jade Plate with Flutist Pattern (Southern Song Dynasty)

Jade plate was the ornament on the belt. In order to distinguish the official ranks, materials of ornaments used on different official costumes have been regulated since the Tang Dynasty. Among them, jade plate was the most distinguished one because of its varied patterns and exquisite workmanship showing the distinct characteristics of every dynasty.

另外，宋代民间工艺品作坊十分兴盛，在北宋都城汴梁（今河南开封）还出现了专门的金钱珠宝首饰店。因此，这一时期的佩饰更是五花八门，品类繁多，且装饰的意义已远远大于实用功能。

辽（907—1125）、金（1115—1234）、元（1206—1368）都是由中国北方少数民族建立的政权。因地域原因，都受到中原汉族的文化影响。元朝统一全国后，接受了南宋文化，同时还受到西域文化的影响。其佩饰虽大体与宋代相近，但有明显的北方草原文化的特征。

In the Song Dynasty folk arts and crafts workshop flourished. Jewelry stores appeared in the capital of the Northern Song Dynasty (now Kaifeng, Henan Province). Therefore, accessories in that period of time were varied. Their ornamental function remained far above the practical use.

The Liao (907-1125), Jin (1115-1234) and Yuan (1206-1368) dynasties were established by the minority in northern China. The culture of Central Plains had impact on the culture of the minority out of the geographical reason. After the unification of the whole country in the Yuan Dynasty, the culture of the Southern Song Dynasty remained. Even if the culture of the Yuan Dynasty was influenced by the Western culture, accessory styles were similar to those in the Song Dynasty but there was no lack of the characteristics of the northern grassland.

舞人形玉佩（辽）

辽代的佩玉受宋代的影响较少，造型不拘一格，神态自然。题材多选自常见的事物，写实性很强。

Jade Pendant with Dancer Pattern (Liao Dynasty)

The style of the jade pendant in the Liao Dynasty had less impact from the Song Dynasty, so jade pendants were not limited to one type and patterns mainly came from common things revealing the realistic feature of life.

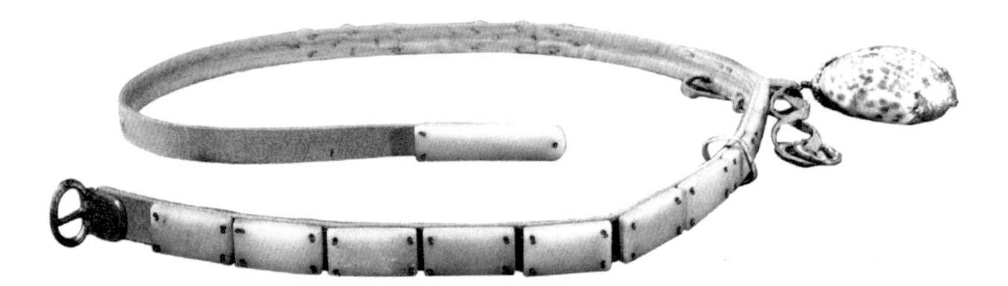

- **玉带（金）**

此玉带由十八块玉带板组成。玉带板每块长11厘米，由金钉连缀在革带上。此玉带还有金扣、金环、玉砣尾，另吊挂一个作为装饰品的贝壳。

Jade Belt (Jin Dynasty)

This jade belt is composed of 18 pieces of jade plates. Each was 11 centimeters long. Gold nails, buttons and rings were used to connect jade plates with the belt. A cowry was hung on the belt for ornamentation.

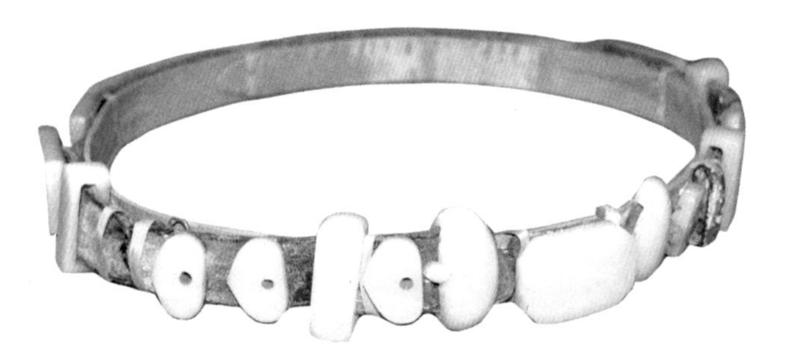

- **玉腰带（元）**

Jade Belt (Yuan Dynasty)

- **龙纹玉带板（元）**

元代玉带板多装饰龙纹、花果纹、鹤纹、狮纹、仕女等，构图丰满，采用近似圆雕的手法刻画形象，立体感很强，刻工精细。

Jade Plate with Dragon Pattern (Yuan Dynasty)

Jade plates in the Yuan Dynasty were mostly decorated with the patterns of dragon, flower, lion, crane and lady at the royal court. This fine carved jade plate using the circular carving technique gave you a strong three-dimensional appeal.

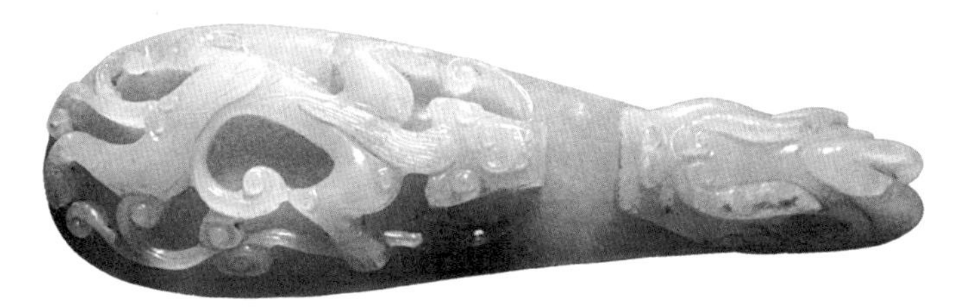

• **龙首螭纹玉带钩（元）**

在宋元时期，人们喜欢把玉带钩束在便服袍外的绦带上，所以玉带钩的数量多了起来。元代玉带钩与宋代玉带钩大体相似，以雕镂精致为特色。

Jade Belt Hook with Dragon Head Pattern (Yuan Dynasty)

People in the Song and Yuan dynasties preferred wearing jade belt hooks on the silk braids outside of the casual gowns. The number of jade hooks increased greatly at that time. Jade belt hooks of the Yuan Dynasty were similar to those of the Song Dynasty. They were characterized by delicate hollow carving.

明清时期（1368—1911）是中国古代佩饰文化发展的鼎盛时期，几乎各种珠宝玉石都可用于制作首饰。尤其是明末清初，缅甸翡翠传入中国，用其制成的首饰和佩件深受人们喜爱。此外，这一时期西洋

The Ming and Qing dynasties (1368-1911) had their heyday in the development of China's ancient accessory. All kinds of gems were used to make accessories. Ever since the emerald was introduced from Myanmar in the late Ming and early Qing dynasties, the emerald accessories and engraving have been loved by Chinese people. On the other hand, glass making technique was also introduced to China resulting in the emergence of glass

• **镂雕蟠螭象牙坠（明）**

Hollow-out Carved Ivory Pendent with Dragon Pattern (Ming Dynasty)

- **双凤纹金链牌（明）**

凤凰是中国古代传说中的百鸟之王，雄性叫“凤”，雌性叫“凰”，通称为“凤凰”。凤凰是一种象征祥瑞的鸟，凤簪寓意生活美满幸福。

Gold Necklace with Phoenix Pendant (Ming Dynasty)

According to a Chinese legend, the phoenix was the king of all birds. It was given the symbolic meaning of happiness. The name of phoenix was the combination of female's name and male's name.

- **龙纹玉带板（明）**

明中期之后，镂雕玉带板十分流行。它多采用多层次镂雕，即在镂雕图案的下面再镂雕一层图案。

Jade Plate with Dragon Pattern (Ming Dynasty)

Hollow-out carved jade plate was very popular after the mid-Ming Dynasty. It adopted the craftsmanship of several layers of hollow-out carvings, that is, the second layer of hollow carving was below the first one.

的玻璃制作技术也传入中国，出现了玻璃材质的佩饰。不但佩饰的造型更为复杂，图案更为繁缛，而且还广泛采用镶嵌宝石、累丝、点翠等技法，使首饰的华丽程度胜过前代。尤其是供明、清王朝宫廷所用的佩饰，造型和纹样等都呈现出雍容华贵的风貌。

在明清时期的佩饰中，具有实用功能的囊、包、袋等占了很大的比重。其余是用金、银、铜、象牙、兽骨等多种材料加工而成的佩件，包括勺、锥、刀、镊、夹等，一般以数件系于腰上或其他部位。

accessories. What's more, because such techniques as inlaid gems and gold wires were widely used, accessories used in the court of the Ming and Qing dynasties were more magnificent than ever before. Diverse shapes and gorgeous patterns could be found on accessories at that time.

Such practical ornaments as bags and pouches accounted for great percentage among all accessories in the Ming and Qing dynasties. Spoons, awls, knives, tweezers and clips were made of different materials such as gold, silver, bronze, ivories and animal bones. They were hung on the waist and other part of the

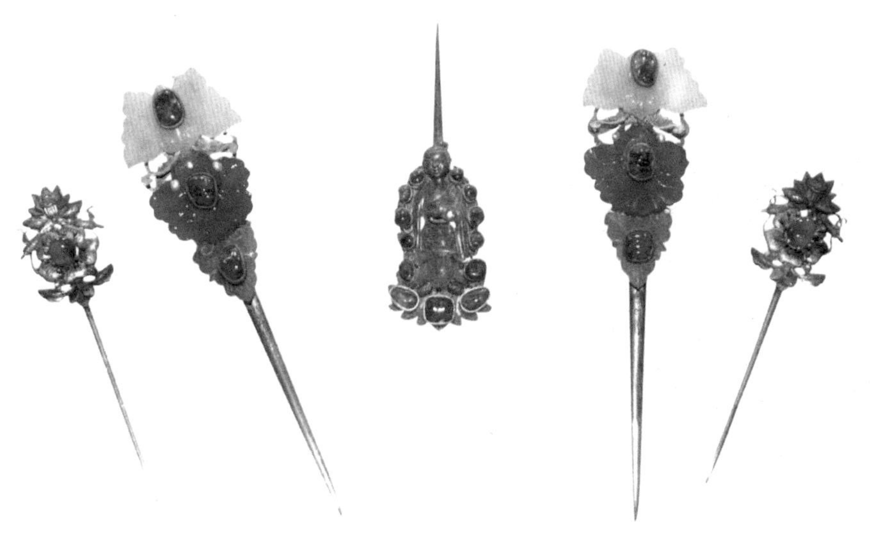

• 镶宝石金簪（明）
Gold Hairpins Inlaid with Gems (Ming Dynasty)

清代中期以后，官宦以及大户人家的男女以在全身佩戴各种物件为时尚。普通百姓身上所佩戴的饰物虽然一般只有几件，但饰物的造型和式样却丰富多彩，不乏别致有趣者。

body for decoration. After the mid-Qing Dynasty, it was fashionable for officials and the rich to wear various accessories all over the body while the common people only wore several pieces but the shapes and patterns were diverse.

• 金嵌珠宝帽顶（清）

此帽饰呈三角形，造型新颖，镂饰缠枝花纹，并嵌有红宝石。

Gold Hat Ornament Inlaid with Jewels (Qing Dynasty)

This hat ornament was in triangle shape. Its novel design showed hollow-out carved interwound branches with an inlaid ruby.

• 拐子纹荷包（清）

Pouch with Decorative Lines (Qing Dynasty)

• **“寿同日月”锁形玉佩（清）**

锁形玉佩是用片状玉制成的锁形的佩饰，表面雕刻吉祥文字，如“长命百岁”“永葆长青”“玉堂富贵”等，一般挂在儿童的胸前。

Lock-shaped Jade Accessory with the Characters of "Longevity with the Sun and the Moon" (Qing Dynasty)

The lock-shaped jade accessory which was hung on the chest of children was engraved with auspicious characters such as "longevity", "evergreen" and "prosperity".

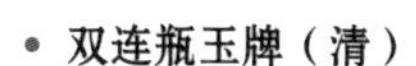

• **双连瓶玉牌（清）**

玉牌也是一种佩饰，多由薄玉板制成，形状有方形、圆形、斧形等。玉牌表面一般雕饰花鸟或人物纹，上部或边部有夔龙纹装饰。

Jade Plate in the Shape of Twin Bottles (Qing Dynasty)

The jade plate was a kind of accessory, which was made into square, round and ax-shaped thin jade plate and decorated with floral or bird patterns and the dragon pattern on the top or on the edge.

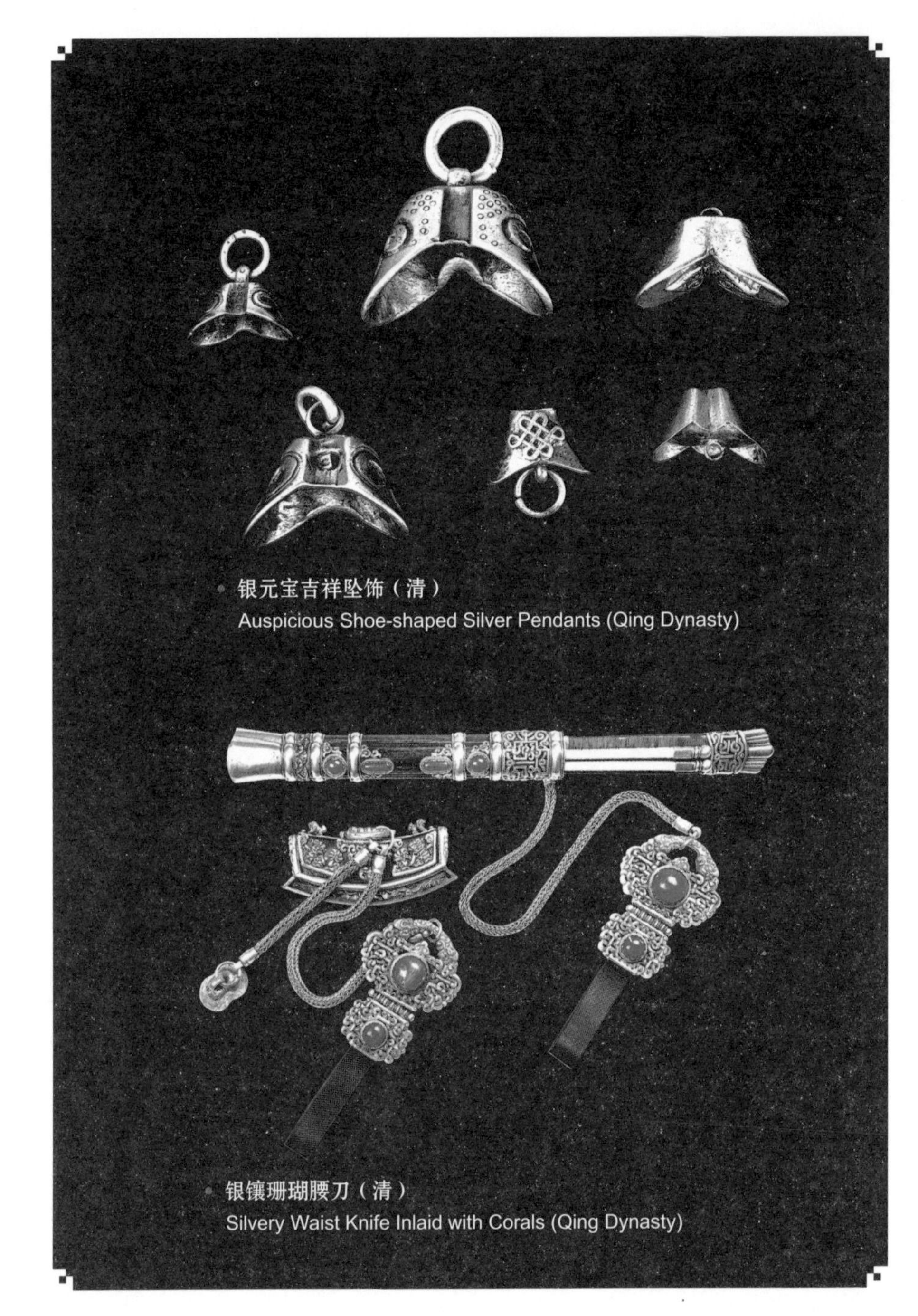

银元宝吉祥坠饰（清）
Auspicious Shoe-shaped Silver Pendants (Qing Dynasty)

银镶珊瑚腰刀（清）
Silvery Waist Knife Inlaid with Corals (Qing Dynasty)

穿金戴银的朝代

元代的贵族喜欢金饰，在服装制作上也大量运用金银饰物。元朝这种“尚金”的风俗超过之前任何朝代，明清时期此风俗也被沿袭。其实，织物加金技术早在春秋战国时期就已出现。但直到东汉时，衣服织金技术也只是在宫廷中使用。魏晋南北朝以后，服饰织金的风气才在全国范围内普及。

- **女金冠（元）**

此金冠由极细的竹丝编结而成，并用藤或竹条制成边圈，再在冠壳表面蒙上麻及薄绢。薄绢用九根金丝箍牢，两侧的金丝弯曲成回旋状。冠的前沿缀有五块镶金边的玉饰。

Gold Crown for Women (Yuan Dynasty)

It was woven with very fine bamboo filaments. Its rim was made of rattans. The surface of the crown was covered with the thin silk or linen and then they were fastened by nine gold wires. The gold wires on both sides were bent into a convoluted shape. Five gold-rimmed jades were decorated on the front rim of the crown.

- **孝端显皇后凤冠（明）**

此凤冠为明神宗孝端显皇后在接受册封、谒庙、朝会时所戴的礼冠。此冠高48.5厘米，重2.320千克。全冠共饰红、蓝宝石一百余块，珍珠五千余颗，金龙、翠凤、珠花、翠叶，金彩交辉，奢华富贵，堪称瑰宝。

Phoenix Coronet for Empress Xiaoduan of the Ming Dynasty

When Empress Xiaoduan of the Ming Dynasty was conferred titles and attended the court ceremony, she wore this coronet. It was 48.5 centimeters high, 2.320 kilograms in weight. Decorated with more than one hundred rubies and sapphires, five thousand pearls, gold dragon, emerald phoenix, bead-made flowers and emerald leaves, dazzling and glorious, the coronet was really a piece of treasure.

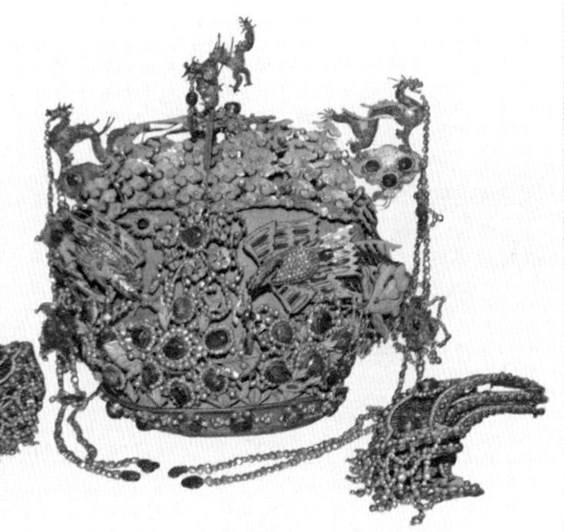

Costume in Gold Thread

Aristocrats in the Yuan Dynasty preferred gold accessories so they were widely used on costumes. The use of gold in the Yuan Dynasty has exceeded any other dynasty. This custom was followed in the Ming and Qing dynasties. In reality, the technique of weaving the gold thread into the cloth appeared as early as the Warring States Period. It wouldn't be used in the royal court until the Eastern Han Dynasty. After the Wei, Jin, Northern and Southern dynasties, the technique was popular nationwide.

- **貂皮嵌珠皇后冬朝冠（清）**

清代皇太后、皇后的朝冠非常华贵，分冬夏两种，冬用熏貂，夏用青绒。冠上覆有红色丝纬，丝纬上缀有七只金凤。冠顶共三层，叠压着三只金凤，金凤之间各有一颗东珠。冠后饰有一只金翟，翟尾垂302颗珍珠，每行饰有青金石、珊瑚等。皇后以下的贵妇的朝冠，以饰品的形制和数目区分。嫔的朝冠饰以金翟，皇子福晋以下以金孔雀装饰。

Mink Fur Court Coronet for Empress Decorated with Beads（Qing Dynasty）

The court coronets for the empress dowager and empress of the Qing Dynasty were very luxurious. They could be fallen into two styles, that is, winter coronet which was made of mink furs and summer coronet which was made of blue velvets. The coronet was covered with red tram silk decorated with seven gold phoenixes. Three overlying phoenixes were put at intervals of a bead. There was a gold pheasant decorated on the back of the coronet hanging 302 pearls with lapis lazuli and corals dotted on every string of pearls. The ranks of aristocrats were distinguished by the shape and the number of ornaments on the court coronet. The court coronet for concubines was decorated with the gold pheasant. Gold peacock decoration was exclusively for the prince's wife and the ranks below.

- **金凤凰饰品（明）**

Gold Phoenix Ornament (Ming Dynasty)

> 古代佩饰的装饰题材与寓意

中国古代佩饰除了具有美化、装饰的功能外，大都还具有吉祥的寓意，佩饰上的图案可谓“图必有意，意必吉祥”。

> Decorative Function and Symbolic Meaning of Ancient Accessories

China's ancient accessories had auspicious meanings in addition to the decorative function, that is, each pattern on them conveyed an auspicious meaning.

There was a variety of decorative patterns in ancient times including figures, animals, flowers, trees, landscapes and so on. People in ancient times adopted homophones and symbolic

• 福禄寿三星绣荷包（清）

福禄寿三星是中国古人为祈求幸福、官禄和长寿而设想的三位神仙。

Embroidered Pouch with Gods of Happiness, Fortune and Longevity Pattern (Qing Dynasty)

In order to pray for happiness, fortune and longevity, the ancient Chinese made up three imaginary gods: god of longevity, god of fortune and god of happiness.

古代佩饰的装饰题材主要包括人物、动物、花草、树木、风景等，并通过谐音、象征等方式，将祈福纳祥、驱恶避害的思想观念含蓄地表达在图案中，然后再通过精湛的工艺表现出来。可以说，每一件佩饰都寄托着人们对美好生活的期盼和追求，都体现着古代工匠们

meanings to endow decorative patterns with implicative meanings of praying for good luck and expelling evils. They were conveyed by exquisite craftsmanship. It could be said that every accessory was full of craftsmen's rich imagination embodying the expectations and pursuit of a better life.

Diverse auspicious patterns could be

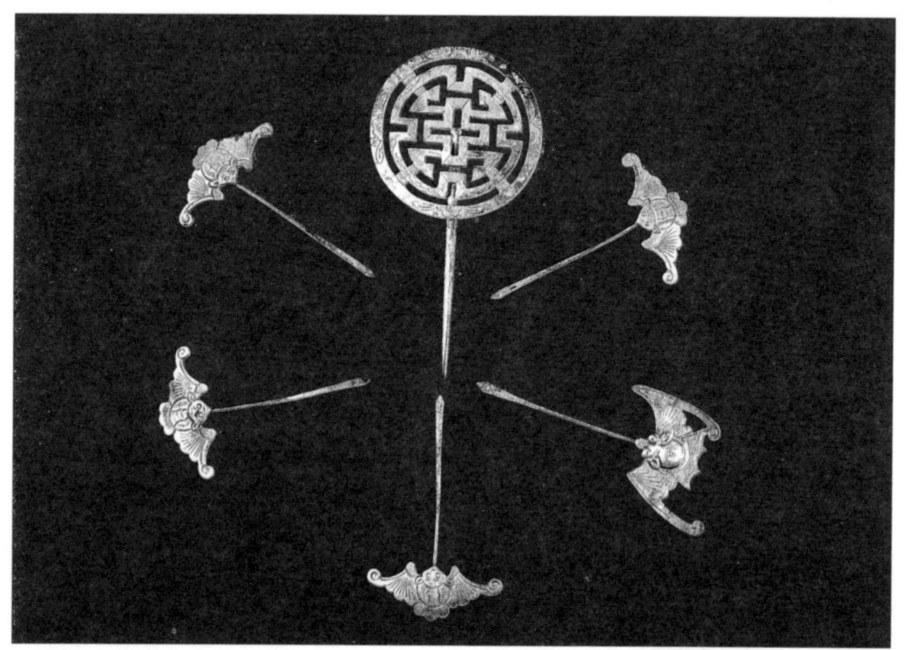

• **银镀金五福捧寿纹簪（清）**

五福捧寿纹的中间是一个“寿” 字，“寿”字周围有五只蝙蝠环绕飞舞。蝙蝠在中国人的心目中是一种吉祥动物，而且“蝠”与“福”同音。此簪表达了祝颂福寿的吉祥寓意。

Gilt Silver Hairpins with the Pattern of Five Bats Circling around Chinese Character "Longevity" (Qing Dynasty)

It showed the pattern of five bats circling around "longevity". Bats are auspicious animals in the eye of the Chinese people. In Chinese, "bat" (*Fu*) is the homonym of "happiness" (*Fu*) conveying the meaning of happiness and longevity.

丰富的想象力和巧妙的构思。

佩饰上的吉祥图案内容十分丰富，常见的就有数十种，如“狮子滚绣球”“聚宝盆”“八仙过海”“五福捧寿”“龙凤呈祥”“双龙戏珠”“连年有余”“麒麟送子”“喜上眉梢”“鲤鱼跳龙门”等。除了这些主体图案之外，云纹、回字纹、万字纹等纹样作为辅助，使佩饰显得更加古朴典雅、生动丰富。

found on accessories, a dozen of which were common to see such as "lions rolling the colored silk balls", "treasure bowls", "the Eight Immortals soaring over the ocean" (it means everyone shows his own skill), "five bats circling around Chinese character of 'longevity'" (meaning happiness and longevity), "dragon and phoenix", "double dragons playing with a bead", "fish" (meaning surplus year after year), "kylin (Chinese unicorn) offering a son" (meaning begetting an heir), "magpies on the prune tree" (it means double happiness is coming), "carps jumping dragon gate" (meaning getting promotion) and so on. In addition to these motifs, cloud patterns and *Huizi* patterns (the shape of the pattern is like the Chinese character " 回 ") as part of the design made accessories more elegant, vivid and colorful.

• 银麒麟送子挂件（清）

麒麟是中国古代传说中的神兽，据说麒麟送来的童子，长大后可成为圣贤之人。因此麒麟被用来象征贵人出世，又成为人们求子嗣的灵兽，在民间深受欢迎。

Silver Pendant with the Pattern of Kylin (Chinese Unicorn) Offering a Son (Qing Dynasty)

Kylin is a sacred beast in Chinese ancient legends. The boy who was offered by Kylin would be a sage. Kylin has been popular among the common people because it has been regarded as an intelligential beast for helping common people beget an heir.

- **翡翠朝珠（清）**

朝珠是清代官员出席正式活动时必备的佩饰。朝珠的大小、质量也表示了官位的高低差别。清朝的官员在觐见皇帝的时候，必须要行“伏地跪拜礼”，且“额头触地”。一些有品级、能够佩戴朝珠的官员，只要朝珠碰地，即可代替“额头触地”。朝珠的直径越大，珠串就越长，佩挂者俯首叩头的幅度就越小。这是皇帝对不同官职的不同恩赐。

Jadite Court Beads (Qing Dynasty)

Court beads were indispensable accessories for officials of the Qing Dynasty to attend official events. The official ranks could be distinguished by the size and quality of court beads. When officials of the Qing Dynasty presented themselves before the emperor, they had to worship the emperor by bowing and kowtowing. The first rank officials and some officials who were entitled to wear court beads didn't have to do that as long as their court beads touched the floor. The bigger the diameter of court beads were, the longer they were. Wearing such beads, the wearers had smaller range of kowtowing, which was the emperor's grace for different ranks of officials.

- **青玉透雕龙形玉佩（战国）**

龙纹是玉器上的主要纹饰之一。龙是古代中国人创造的一种虚拟动物，乃万兽之首。传说龙身长若蛇，有鳞似鱼，有角似鹿，有爪似鹰，能走能飞，能大能小，能隐能现，甚至能翻江倒海、吞风吐雾、兴云降雨。在中国古代社会，龙是帝王的象征，凡是帝王的器物，大都以龙为纹样。

Gray Jade Pendant in Dragon Shape (Warring States Period)

The dragon pattern was one of the main ornamentations in successive dynasties. In legends, a dragon has a snake-like body, fish scales, antlers and hawk talons. The dragon is a fictitious creature created by ancient Chinese. As the head of all beasts, it has the capability of overturning rivers and seas, calling up the wind and invoking the rain. In ancient times, the dragon emblematizes the emperor, and any utensils used by emperors have the pattern of dragon.

经典佩饰
Classical Ancient Accessories

中国古代的佩饰种类繁多，主要包括佩玉、带钩、荷包、香囊、长命锁、鼻烟壶、压胜钱等，此外还包括首饰。这些佩饰造型精美、工艺精湛，体现了古人祈求平安吉祥的愿望。

There were a wide range of accessories in ancient times mainly including jade accessories, belt hooks, pouches, sachets, longevity locks, snuff bottles and auspicious coins. These exquisite accessories reflected the ancient people's common wish for safety and propitiousness.

> 佩玉

"古之君子必佩玉"，玉在中国传统文化里有着不可替代的作用。在原始社会，玉被视为山川精华，成为神灵的象征之一。人们不但把玉作为图腾崇拜，还把带有某种含义的玉石佩饰作为氏族的标志。玉被雕琢成鸟、兽、神人等造型，或在玉器表面绘制神人、兽面等图案。由于玉的数量稀少而且加工困难，只有族群里的族长、祭司等才有资格佩戴并使用它。因此，玉佩饰成为权力、地位、神权的象征。

夏朝统一天下后，玉成为国家政治的重要内容，被用于区别等级关系，或用于祭祀。

春秋战国时期，古人将玉在质地、光泽、硬度、纹理、音色等方

> Jade Accessories

As a saying goes, "Every gentleman in ancient times has a jade with him." It would be safe to say that jade played an irreplaceable role in traditional Chinese culture. In primitive society, jade as a symbol of gods, was regarded as essence of mountains and rivers. Ancient people not only worshipped jade as their totem but also took jade with symbolic meaning as the sign of the clan. Different patterns such as birds, beasts, gods and figures were engraved on jade. The material was scarce and patterns on jade were hard to carve at that time. Only the head of the clan and the priest were entitled to wear jade accessories. Jade accessories symbolized the power, social status and theocrat accordingly.

After China achieved unification in the Xia Dynasty, jade played an important role in state administration such as

七璜联珠组玉佩（西周）

此组佩是西周重要的礼器，分为上、下两部分，上部由玛瑙珠、玉管组合成项饰，下部用玉璜、玛瑙珠和料珠组合。

Seven Jade *Huang* in a Group (Western Zhou Dynasty)

In the Western Zhou Dynasty, it was an important sacrificial jade accessory which was connected by agate beads and tube-shaped jade on the upper part of the accessory and jade *Huang*, agates and other beads on the lower part.

面的优良品质，与个人的道德规范和品德修养联系在一起。因此佩玉的传统一直流传下来，玉被视为君子的象征。佩玉就是要时刻提醒着君子要牢记玉的品德，并以此匡正自己的行为。

玉器在相当长的一段历史时期内，只供上层社会使用，平民百姓很少能接触到玉器。隋唐时期随着生产力的不断发展，玉开始脱离礼制和阶级属性的束缚，走向民间，

distinguishing hierarchical relationship, stabilizing political situation and offering sacrifices to gods and ancestors.

In the Spring and Autumn Period people started to link the quality of jade such as its texture, luster, harness, grain and sound with the personal ethic and morals. Thus, the tradition of wearing jade accessories has been passed down. Jade was seen as the symbol of gentleman, which served as a reminder to the wearer keeping the essence of jade in mind so as to remedy his behavior.

Because jade had been for the upper-class people's exclusive use for a long period of time, common people rarely had jade accessories. With the development of productive forces in the Sui and Tang dynasties, when jade accessories unshackle the restriction of feudal ethics and rites, such jade parts as bowls, cups, hairpins and bracelets were

出现了玉制的碗、杯、簪、镯等容器和饰品。明清时期，玉雕技术逐渐成熟，各地的玉作坊由此兴盛起来。玉佩饰的制作工艺越来越精细，题材和种类也不断扩展，运用人物、动物、花鸟、器物，以及一些民间谚语、吉语及神话故事，通过借喻、比拟、双关、象征、谐音等表现手法，构成“一句吉语，一幅图案”的表现形式。玉佩饰成为吉祥、美好的象征。

总之，在中国古人看来，玉是吉祥的圣物，有着神秘莫测的灵气，一块上好的玉件能给人带来好运，使人趋吉避凶。同时，玉也是

used by common people. The technique of jade carving was well developed in the Ming and Qing dynasties resulting in workshops flourishing in every corner of China. Craftsmanship of jade carving became all the more sophisticated with the increasing styles and patterns. Making full use of such technique of expression as metaphor, analogy, pun, harmonic tone and symbolic meaning, figures, animals, flowers, birds, vessels, images in the idioms and myths were endowed with auspicious meanings.

In short, in ancient Chinese's eyes, jade was taken as an auspicious and scared object which could bring good luck for wearers. On the other hand, jade

• **鸟首人身玉佩（商）**

此玉佩呈片状，鸟身较短，身饰双阴线折线纹，高冠，冠上饰凸齿，钩喙；鸟下身如跪地人身，圆臀，腿置于臀下。这种鸟首人身的玉佩是时代的产物，缘于人对神与动物图腾的混合崇拜。

Jade Pendant with the Combined Pattern of Bird and Man (Shang Dynasty)

It was made of a thin piece of jade. The body of the bird was short decorated with double intaglio lines. Convex teeth outlined its pileums and beak. There seemed a man kneeling down on the lower part of the bird. His legs placed under his round hips. The combined pattern was the product of that period when the mixed totem of gods and animals was worshiped.

搭配在服装上的点睛之笔。在古人看来，玉的装饰作用是任何金银珠宝饰品都无法替代的。

佩玉又叫“玉佩”，主要可分为两种：一种是用薄片玉板制成的单件佩玉，单件佩玉一般按纹饰命名；另一种是用多件玉饰有序组合而成的组佩。古代的玉佩还有大佩、组佩之分，以及礼制以外的装饰性玉佩。

大佩，是单件玉佩中最为重要的一种，属于礼器。使用时将其悬

accessories would add colors to wearers. Any other jewelry was inferior to jade accessories.

There were two kinds of jade. One was made of one-piece jade plate which was generally named after its pattern. The other was made of combination of multi-piece jade. In addition to *Dapei* (a kind of single piece pendant) and groups of jade pendants, there were other types of ornamental pendants which had no relation with feudal ethics and rites.

Dapei, a kind of jade sacrificial

- **青玉虎形佩（春秋）**

虎形佩常成对出土，一般由薄片玉板制成。此虎形佩形似老虎，虎头较大，近似方形，上唇大于下唇，上下唇皆上卷，呈方形。虎背由两道连弧线构成，尾粗而大，自臀部垂下，尾端上卷。虎身布满饰纹。

Tiger-shaped Jade Pendant (Spring and Autumn Period)

It was made of a thin jade plate outlined a crouching tiger. The tiger's head was big enough to be in square shape. Its square-shaped upper lip was bigger than the lower one. Both lips were curved upward. Double arcs outlined the back of the tiger. Its long and thick tail curled upward in oblong shape. The tiger's body was ornamented with decorative patterns.

• **龙凤玉组佩（战国）**

此组佩出土于湖北曾侯乙墓，通长48.5厘米，最宽处为8.5厘米。玉为青白色，呈长带状，五块玉料共雕成十六节，椭圆形的活环和榫把它们连接在一起。各节分别透雕成龙、凤或璧形。璧上主要饰谷纹，间或有云纹和斜线纹。龙、凤则以阴刻和浅浮雕表现出嘴、眼、角、鳞甲、羽毛、尾、爪。全器透雕、浮雕和阴刻共刻出37条龙、7只凤和10条蛇。

Groups of Jade Pendants with Dragon and Phoenix Patterns (Warring States Period)

The jade was 48.5 centimeters in length and the max width was 8.5 centimeters unearthed in the Tomb of Marquis Yi of the Zeng State, Hubei Province. It was bluish white in strip shape. 16 parts were engraved on 5 pieces of jade and they were tenoned by elliptic rings. With techniques of openwork carving, intaglio and relief, 37 dragons, 7 phoenixes and 10 snakes were engraved on the group of jade pendants. Intaglio and relief were used to present the details such as mouths, eyes, antlers, hawk talons and feathers. The cereal pattern as the main ornamentation together with the cloud pattern was engraved on semi-annular jade pendants.

挂在腰下，专门用于贵族男女的祭服或朝服。这种制度始于商周，历代沿袭，入清之后被废止。

组佩也称“杂佩”，由玉珩、玉璜、玉琚、冲牙等多种玉器组合而成。玉组佩在商周至两汉时期非常流行，为王公贵族必佩之物。根据佩带者身份地位的高低，组佩

vessel, was one of the important single piece pendants hung on the waist. It was for noble men and women's exclusive use on their ceremonial robe and court costume beginning with the Shang Dynasty and *Dapei* was not abolished until the Qing Dynasty.

Groups of jade pendants were combined with different shapes of jade. It was very popular among princes and aristocrats in the Shang, Zhou, Eastern Han and Western Han dynasties. The wearers' social status varied in the size, shape and structure of groups of pendants. After the Han Dynasty, the institution

的大小、结构都不相同。汉代以后，玉组佩逐渐被废止，明代时再度流行，成为冠服制度中不可缺少的佩饰。

除此之外，装饰性的玉佩包括生肖形玉佩、人纹佩（如武士形象佩）、龙纹佩、凤纹佩、禽纹佩等，这类玉佩往往比大佩更加细腻、逼真和精美。

related to wearing groups of pendants was abolished but it resurged in the Ming Dynasty and became an indispensable part in the costume and crown institution.

In addition, there were decorative jade pendants with different patterns such as chinese zodiac figures, dragons, phoenixes, and poultry. They were more delicate and lifelike than *Dapei*.

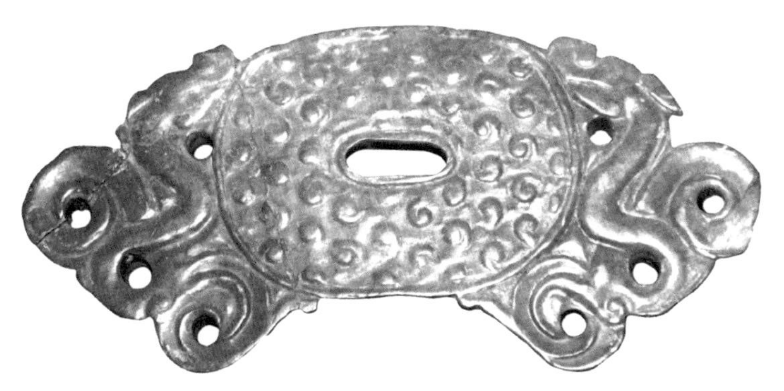

• **双夔耳璧形玉佩（战国）**

夔龙是中国古代传说中的一种神兽，只有一足，类似爬虫。

Jade Pendant in *Bi* Shape with Two *Kui* Ears (Warring States Period)

Kui was a one-legged magic beast which looked like a reptile in Chinese fable.

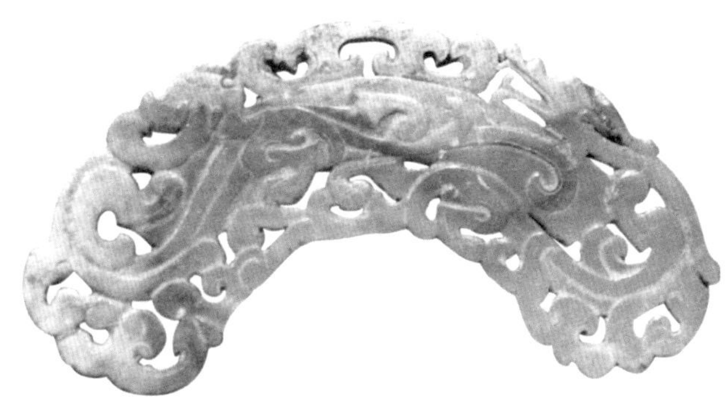

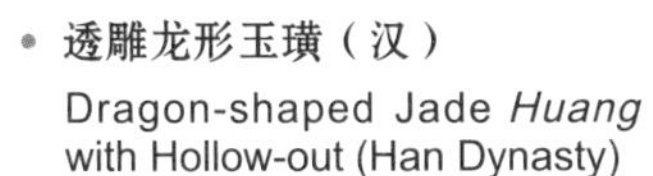

• **透雕龙形玉璜（汉）**

Dragon-shaped Jade *Huang* with Hollow-out (Han Dynasty)

- **装饰性玉佩（宋）**

在中国古代，对于装饰性玉佩的使用也是有等级规定的，具有显示佩戴者身份的作用。

Decorative Jade Pendants (Song Dynasty)

In ancient China, the wearers' social status could be identified based on the decorative jade pendants they worn.

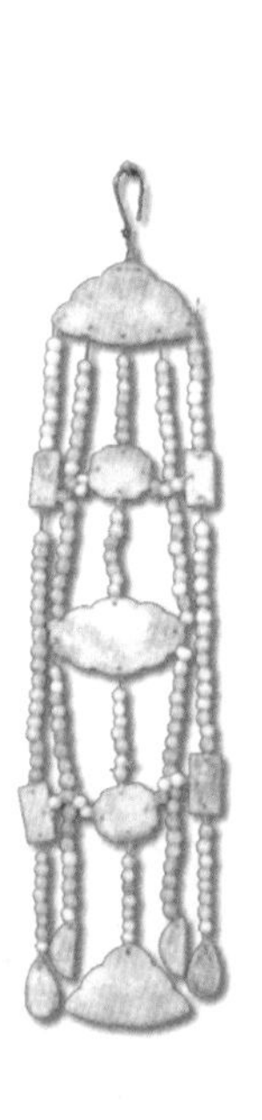

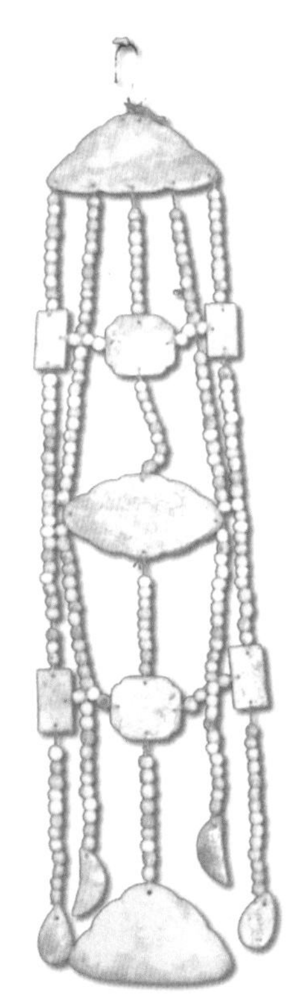

- **北京定陵出土的玉组佩（明）**

明代墓葬中出土了多套玉组佩。这套玉组佩所用的白玉质地优良、做工精致，是御用珍品。

Groups of Jade Pendants Unearthed from Beijing Dingling Mausoleum (the Tomb of Emperor Wanli of the Ming Dynasty and His Two Empresses)

Among a number of unearthed groups of jade pendants, this exquisite jade accessory which was made of high quality white jade was exclusively for the royal palace.

- **盘长纹玉佩（清）**

中国几千年来爱玉之风盛行。在有些场合，无玉便不能参加集会。此玉佩以盘长纹为主体，做工精细，玉质滑润。盘长纹盘曲连接，无头无尾，无终无止，寓意源远流长、无限生机，有连绵不断、子孙延绵之意。

Jade Pendant with Twining and Curving Pattern (Qing Dynasty)

The past several thousand years witnessed the popularity of jade accessories. People had to wear jade accessories when meeting with their friends in some situations. This jade pendant was made delicately. Its pattern featured endless twining and curving without beginning or ending which expressed the meaning of everlasting vitality and fertility.

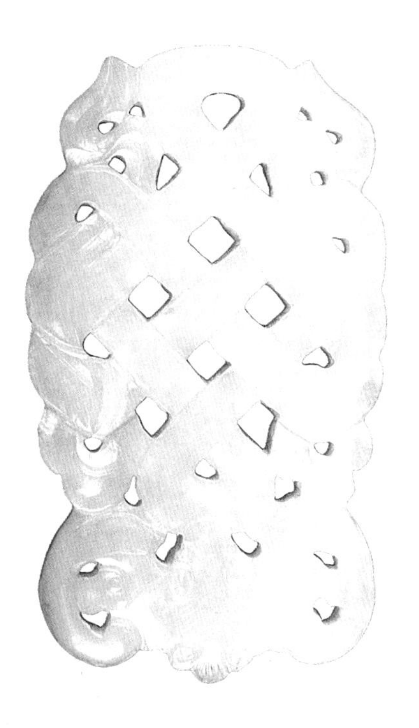

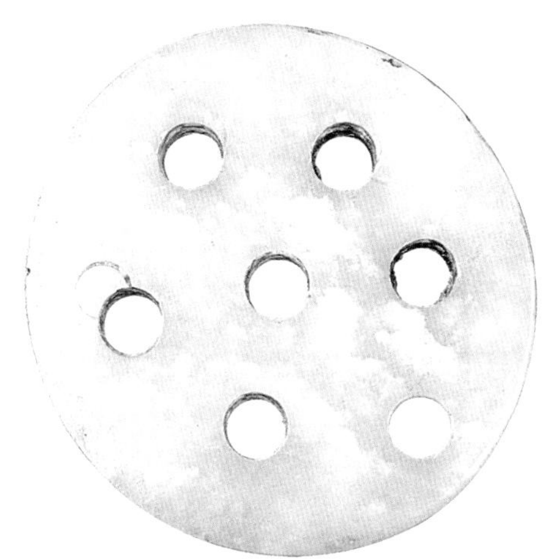

- **路路通玉佩（清）**

此玉佩形状极似莲藕的横断面，因此又称“藕断丝连”，多为男女之间的定情物。“藕片”上面的洞被称为“路路通”，寓意路子多，哪条路都通。

Jade Pendant with Lotus Root Cross-section Pattern(Qing Dynasty)

Its pattern looked like a cross-section of lotus root. It symbolized the love between men and women, like a lotus root, when it was divided, fiber remained connected; therefore, this jade pendant was mostly used as a token of love. There were several holes on the pendant symbolizing "lots of accessible paths", well-connected social and business network.

“行步则有环佩之声”

“行步则有环佩之声”出自《礼记·经解》。“环佩”本义为中国古人所佩戴的佩玉，后来则多指年轻女子所佩戴的玉饰，“环佩”也渐渐成了女性的代称。

中国古代女子的佩玉，常以丝线贯串，结成花珠形制，更以珠玉、宝石、钟铃来搭配，通常系在衣带上。因此，佩戴环佩的女子走起路来会发出“叮当”的悦耳声音。环佩的一个重要的作用就是提示女子要注意自己贤淑的仪态，当佩戴者的动作稍微大些，环佩就会发出声音，借以提示佩戴者要小心些，否则就会出丑了。

"Jade Pendants Jingle at Every Step" (a Quote from *The Book of Rites*)

The Jade pendant worn on a girdle was the general term of the jade accessory worn by ancient Chinese. Later, it referred to the jade accessory worn by women in particular and became the pronoun of women accordingly.

Ancient women's jade pendants were linked up with silk strings, beads, gems and bells in floral shape. They tied pendants on the girdle. When they walked, jade pendants were jingling at every step. Jade pendants were used to remind women of minding their virtuous manners. The wearers had to walk gracefully in case jade pendants would make big sound.

- **《美人图》中的清代女子**

此清代女子体态婀娜，腰间佩玉。

***The Painting of Beauty* Depicted a Woman of the Qing Dynasty**

She looked elegant with jade pendants on the waist.

> 带钩

中国古人的腰带有大带和革带之分。大带以布帛制作，用于束腰紧身，而革带主要用于系佩玉、印章、香囊、刀剑等物品。因革带硬而厚实，无法同大带一样系结，所以在使用时多借助于带头。此类带头通常被制成钩状，即“带钩”。

> Belt Hooks

Ancient Chinese commonly wore two types of belt. One was made of leather which was used to tie jade pendants, seals, pouches, and swords. The material of leather belt was so stiff that the buckle had to be used on the leather belt. The other was made of textile, which was used to fasten the costume. They were

• 金带钩（春秋）
Gold Belt Hook (Spring and Autumn Period)

• **鎏金铜带钩（战国）**
此带钩为琵琶形，工艺精美，制作十分考究。
Gilt Copper Belt Hook (Warring States Period)
It was made exquisitely in *Pipa* (4-stringed Chinese lute) shape.

带钩一般为贵族和文人、武士所用，是身份的象征。战国至秦汉最为流行，多由青铜、玉、金、银、铁等制成。早期的带钩主要是实用品，后来发展成具有实用性的高贵装饰品。

带钩有着悠久的历史，曾出土过大量的新石器时代的玉带钩。商周时期开始用名贵的美玉制作带钩。

春秋早期，普通百姓开始把铜带钩固定在革带的一端。只要把带钩钩住革带另一端的环或孔眼，就能把革带钩住，既方便又美观，因而被大量使用。到了战国时期，带

usually made in hook shape, therefore called "belt hook".

In the Warring States Period and the Qin Dynasty, belt hooks were generally used by the aristocracy and warriors to show their distinguished social status. They were made of bronze, jade, gold, silver, iron and other materials. In early times, they were used to tie the clothes but later they became the luxurious ornaments besides the practical use.

The history of wearing jade hooks could be traced back to the Neolithic Age based on a large number of unearthed hooks. Starting from the Shang and Zhou dynasties, belt hooks were made of precious jade.

In the early Spring and Autumn Period, ordinary people began to fix the cooper hook on one end of the belt to buckle the other end of the belt. It was widely used. When entering the Warring States Period, the belt hook became

钩变得越发华丽起来，除了选材考究之外，制作的工艺也更为精湛。战国时期的带钩形式多种多样，但钩体都呈“S”形，下面有柱。春秋战国时期的带钩材料主要有玉、金、银、青铜和铁。带钩的制作工艺除雕镂花纹外，还有镶嵌、鎏金等工艺。人们还会在带钩上镶嵌一些绿松石、猫眼石等。

西汉时期的带钩以玉制的为主，带钩的制作在继承前代形状和技法的基础上又得到了进一步的发展和创新。西汉时期的玉带钩大多通体光素无纹，有的甚至仅为轮廓，有的只是刻画几刀而已，但都粗犷有力、规整洁净。不过也有精雕细琢的玉带钩，并开始出现了浅

even more gorgeous. It was featured with well-selected materials and delicate workmanship. The main structure was in an "S" shape with dozens of variants. Materials of the belt hook in the Warring States Period were jade, gold, silver, bronze and iron. In addition to hollow-out carving and gilding, inlay technique was also used, therefore the hook was inlaid with turquoise and opal at that time.

The hook in the Western Han Dynasty was mostly made of jade. Carried on the techniques in previous dynasties, the jade hook made further development and innovation. Patterns on some of jade hooks were engraved simply and roughly and some were even without any pattern but there were exquisitely carved jade hooks. Bas-reliefs were used

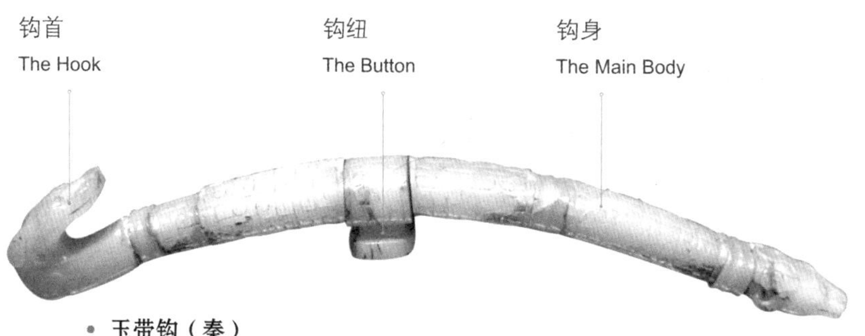

- **玉带钩（秦）**

带钩由钩首、钩身和钩纽三部分组成。

Jade Belt Hook (Qin Dynasty)

It was composed of the hook, the main body and the button.

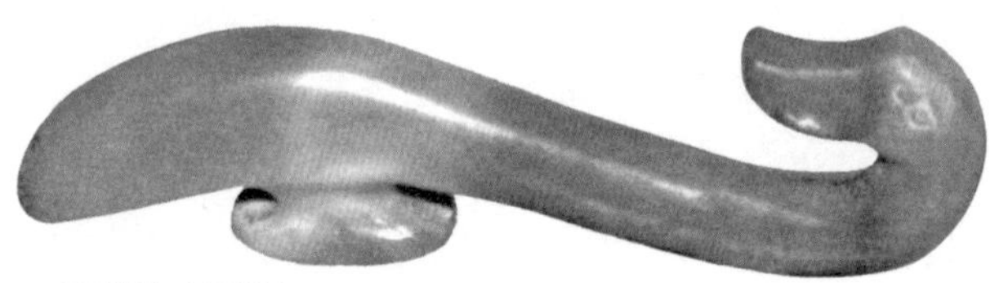

玉带钩（西汉）

西汉时期的玉带钩造型简单、挺拔、有力。

Jade Belt Hook (Western Han Dynasty)

The structure of the jade belt was simple but stout.

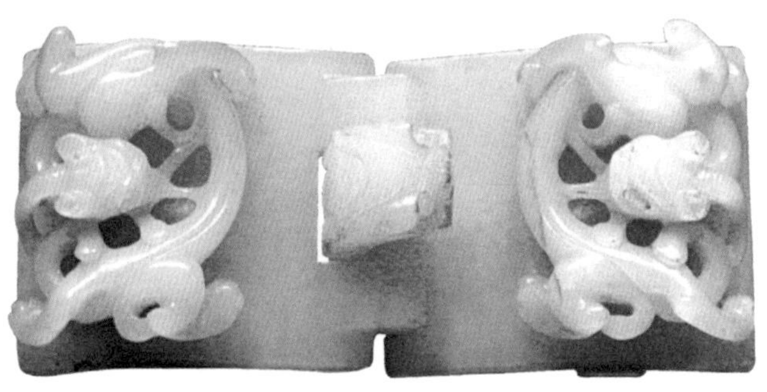

白玉螭虎纹龙首带钩（明）

明清时期的玉带钩上多有浮雕纹饰，所雕兽纹凸出钩面，呈腾空之势。此带钩由分离的两部分组成，一部分是伸出的螭首钩，一部分是突出的承受钩的扣环，因此又名“玉带扣”。

White Jade Belt Buckle with Beast Heads Pattern (Ming Dynasty)

Relief was widely used on jade buckles in the Ming and Qing dynasties. Beasts were on the front side of the belt buckle as if they were soaring. It was composed of two parts: one was a cramp ring and the other was a hook, hence known as jade belt buckle.

青玉嵌宝石带钩（明）

此带钩为如意形，如意是中国古代一种具有吉祥寓意的物件，形似灵芝，寓意平安如意、福寿如意、万事如意等。此带钩的钩首雕成螭首，螭首转折后变为玉带钩的板状钩体，板状钩体的背面有一个大圆纽。此带钩体形厚重，制作十分精美，镶嵌有各色宝石，极具赏玩价值。

Gray Jade Belt Hook Inlaid with Gems (Ming Dynasty)

The shape of the hook was a *Ruyi* (S shaped ornamental object, a symbol of good luck) inlaid with colored gems. The beast head was used as the hook. There was a big button beneath. It looked dignified.

浮雕的蟠螭、凤鸟等纹饰。

东汉以后，带钩的数量锐减，造型也相对单调。

元明清三代是带钩制作的又一鼎盛时期，出土和传世的数量很多，并且大都造型优美、技艺高超。但此时的带钩已由实用品变成了装饰品。这一时期的带钩一般都有花草、动物等纹饰，钩头多为龙头形，尤以龙螭纹相组合的龙带钩最为精美。

on the jade hook such as patterns of the legendary dragon and phoenix.

After the Eastern Han Dynasty, the number of the jade hooks dropped greatly and the style was relatively dull.

A great number of unearthed belt hooks proved that the jade hook production reached another heyday in the Yuan, Ming and Qing dynasties. Such patterns as flower, grass and animal were engraved exquisitely, among which the dragon head shaped hook was especially made sophisticatedly.

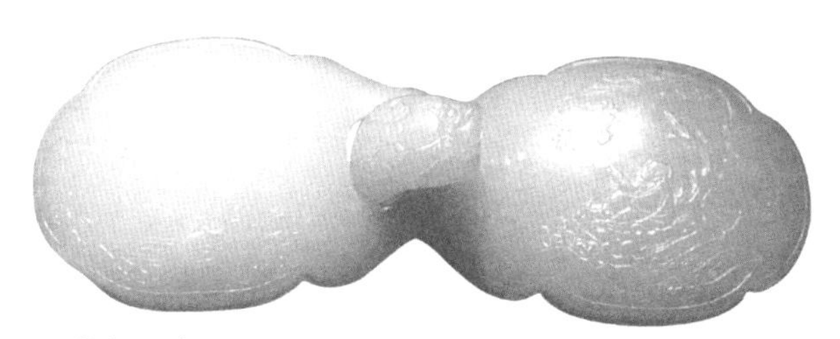

- **玉带扣（清）**

玉带扣是清代常见的品种，一般为椭圆形，没有尖锐的硬角或突起，器表有极浅的浮雕纹饰。

Jade Belt Hook (Qing Dynasty)

It was a common style in the Qing Dynasty. There were no sharp corner and protrusion on the jade belt buckle so as to avoid ripping the clothes. It was oval in shape decorated with bas-reliefs.

- **青玉镂雕蟠螭纹龙首带钩（清）**

此龙首带钩浮雕精美，钩首的龙头微微昂起，极具神韵。

Hollow-out Carved Gray Jade Belt Hook with Dragon Pattern (Qing Dynasty)

The vivid relief showed that the dragon perked up its head slightly, which was used as a buckle.

> 荷包、香囊

在中国古代，人们的衣服上并没有设计衣兜等存放散碎物品的装置。但古人会随身佩带一种可以装铜钱、金银、票据、印章、手帕、针线等散碎物品的小包，即荷包。

“荷包”一词最早出现在宋代，发展到清代时，荷包的造型已丰富多样。荷包的制作材料有很多种，包括丝绸、麻布、棉布、毛皮、金帛等。荷包的造型也是多种多样，最主要的有圆形、椭圆形、正方形、长方形、心形、如意形、石榴形等。荷包的图案亦非常丰富，几乎每个荷包都不尽相同，花卉、鸟兽、草虫、山水、人物、诗词歌赋等都可以成为荷包上的图案。有大量的清代荷包传世，通常以丝织物做成，上施彩绣。

> Pouches and Sachets

There was no pocket on the ancient garment to hold such necessities as money, seal, handkerchief, voucher, thread and needle. To solve this problem, ancient people designed pouches.

The word "pouch" was first used in the Song Dynasty. In the Qing Dynasty, pouches were diverse in the style, material and shape. They were made of silk, linen, cotton, fur and golden silk. The pouch was round, circular, oval and rectangular in shape. Other shapes of the pouch included heart, *Ruyi* and pomegranate. There were various patterns on the pouch which was distinct from one another. They included flowers, birds, beasts, grass, insects, figures and poets. A great number of pouches passed down from the Qing Dynasty. Most of them were made of silk fabrics embroidered with colored thread.

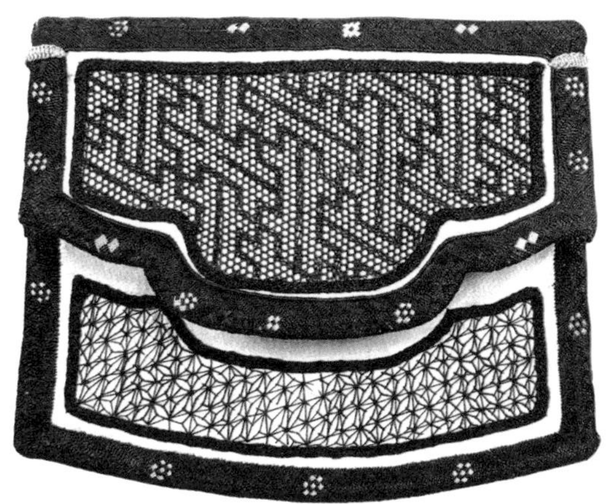

- **网绣万字纹荷包（清）**
  网绣是将丝线组成网状结构的刺绣技法。万字纹是一种常见的吉祥纹样，寓意吉祥如意和荣华富贵。
  Embroidered Pouch (Qing Dynasty)
  Net-like pattern was made by a special embroidery skill.

香囊与荷包类似，但相对于荷包较强的实用性，其装饰性更强一些，且主要用于贮放香料。早在先秦时期，人们就有佩挂香囊的风俗了。当时每到端午时节，一些儿童就会佩戴香囊。

宫廷或富贵人家的一些香囊常用金银制造，佩挂在身上或悬挂于床帐内。而大多数的民间香囊则多由透气性好的素罗制成，上面绣上鸳鸯、莲花、梅花、菊花、桃子、苹果、娃娃骑鱼、娃娃抱公鸡等寓意吉祥的图案。香囊里填充的香料或药材，主要有艾草、苍术、白芷、菖蒲、藿香、佩兰、川芎、香附、薄荷、香橼、辛夷等，此外还可以适当加入苏合香、益智仁、高

Compared with pouches' practicality, sachets for perfume storage tended to be more decorative. As early as the Pre-Qin Dynasty, people had the custom of wearing sachets. And children wore sachets on the occasion of Dragon Boat Festival.

Gold and silver sachets were used by the wealthy and the royal palace while most of sachets used among common people were made of plain gauze with embroidered mandarin ducks, lotuses, plum blossoms, chrysanthemums, peaches, apples and auspicious patterns such as "a baby ridding a fish", "a baby holding a rooster". They were hung on the bed curtain. All kinds of traditional Chinese medicine and spices were filled up sachets. They included wormwood,

良姜、陈皮、零陵香等药材。

香囊与荷包除了作为佩饰以外，还是中国古代许多青年男女的定情信物。女子常将自己亲手绣制的香囊或荷包赠予男方，以表达自己的爱慕与牵挂。

angelica, calamus, agastache rugosus, herba eupatorii, Ligusticum wallichii, Cyperus, mint, citron, Flos Magnoliae, storax, semen amomi amari, orange peel and holy basil, etc.

Sachets and pouches were also seen as a token of love in addition to being used as accessories. The girl often gave her embroidered sachet or pouch to her beloved expressing her affections.

• 褡裢（清）

褡裢是中国古代长期使用的一种袋囊，布制长形，有两个口袋，可盛装物品，出门时往肩上一搭。

Dalian (Qing Dynasty)

Ancient Chinese wore the bag which was made of long-shaped cloth with two pockets for holding something, called *Dalian*. When they went out, they carried it on the shoulder.

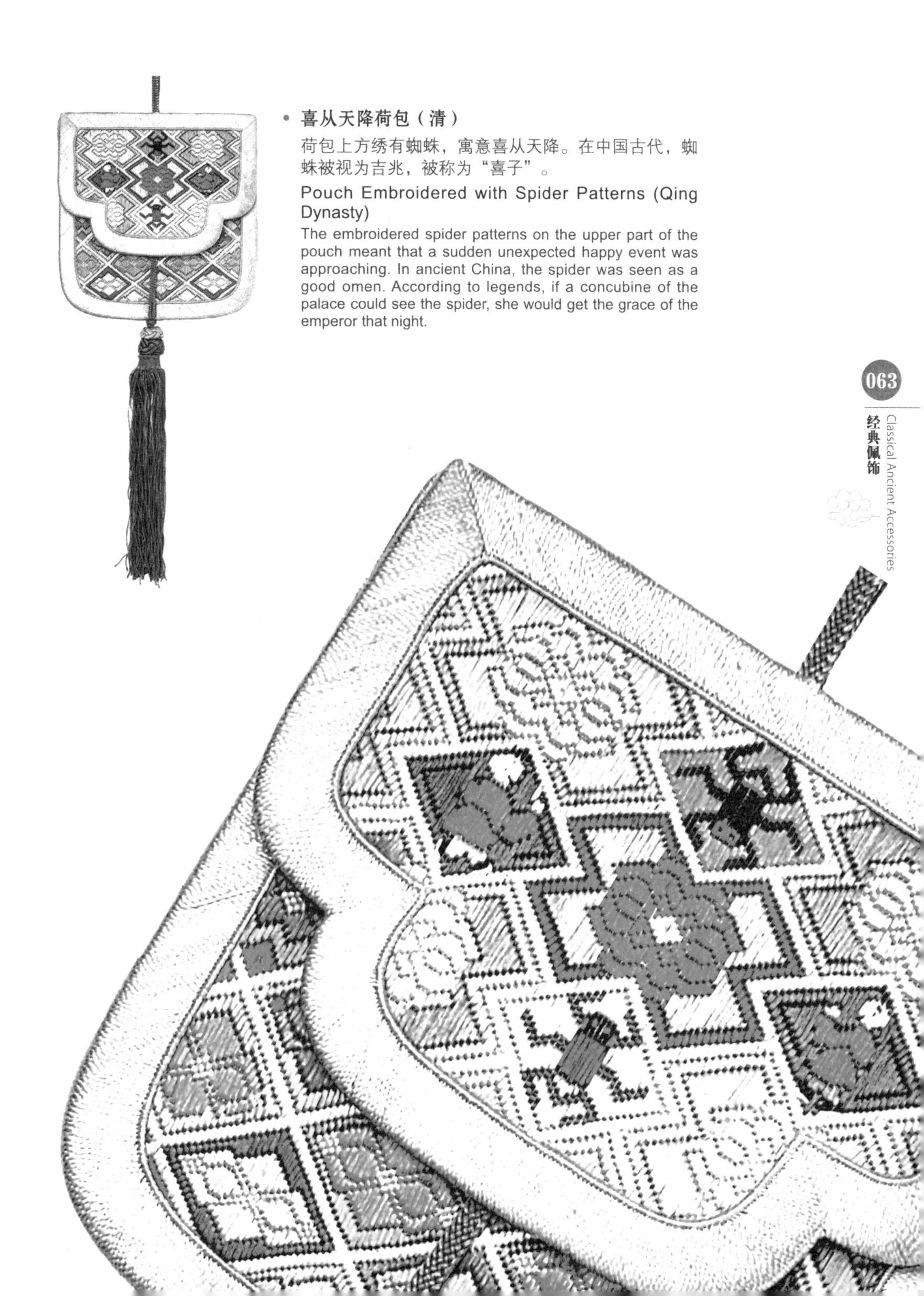

喜从天降荷包（清）

荷包上方绣有蜘蛛，寓意喜从天降。在中国古代，蜘蛛被视为吉兆，被称为“喜子”。

Pouch Embroidered with Spider Patterns (Qing Dynasty)

The embroidered spider patterns on the upper part of the pouch meant that a sudden unexpected happy event was approaching. In ancient China, the spider was seen as a good omen. According to legends, if a concubine of the palace could see the spider, she would get the grace of the emperor that night.

"三娘教子"荷包（清）

"三娘教子"是一个民间故事。讲述的是明代一位商人娶了三个妻子，后来在外出经商时商人不幸身亡。其大妻、二妻相继改嫁，只有第三位妻子含辛茹苦地将大娘所生之子倚哥抚养长大。但倚哥长大后，知道三娘不是自己的亲娘，便不服三娘管教，不肯学习。三娘气愤之下，砍断织机，并用剪刀把织好的布剪断，教育倚哥："你荒废学业，就如同我剪断这布一样，将一事无成。"从此以后，倚哥苦读诗书，最终金榜题名。

Embroidered Pouch with the Pattern of "Sanniang Teaching Her Son" (Qing Dynasty)

"Sanniang teaching her son" is an ancient Chinese folklore. The story happened to a businessman in the Ming Dynasty. After he was killed in an accident when on business, his first and second wives abandoned the family and remarried, only his third wife (Sanniang) endured all kinds of hardships to bring up Yige (the son of the first wife). When Yige grew up, he got to know that Sanniang was not his biological mother. He was unwilling to study and did not follow Sanniang's discipline any more. Irritated by Yige, she broke the loom and cut off all cloth to educate Yige, "If you abandoned your studies, you would achieve nothing just as I cut off all woven cloth." Since then, he has studied so hard that he succeeded in the government examination finally.

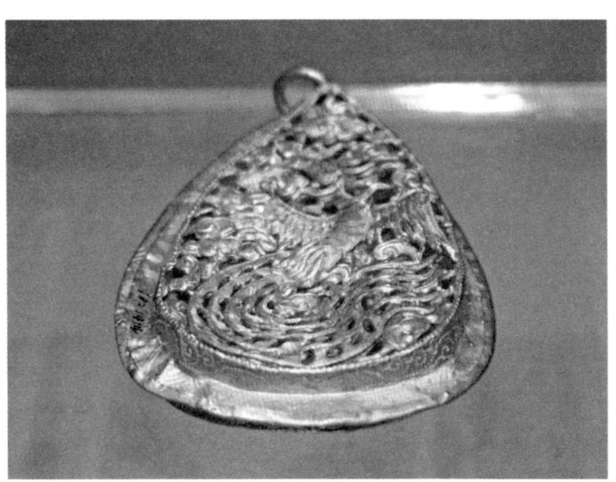

金香囊（明）

此金香囊上有金钩，下面的轮廓呈鸡心形，双面透雕有野雉飞翔纹样，出土时里面有一个丝质囊袋。

Gold Sachet (Ming Dynasty)

There was a hook on the top of the gold heart-shaped sachet. The pattern of the openwork carved flying wild pheasant was on both sides. When it was unearthed, there was a silk pouch inside the sachet.

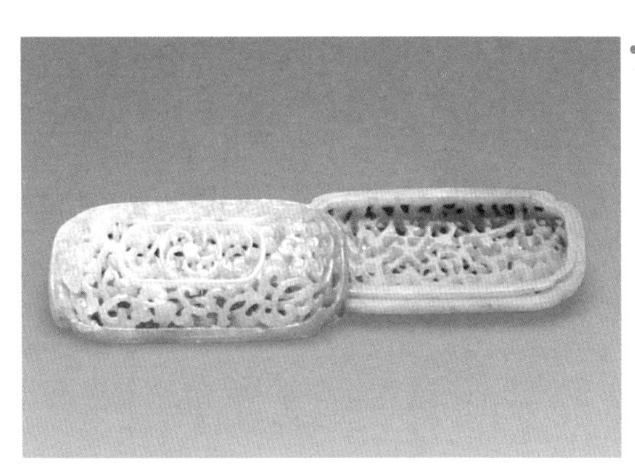

青玉勾莲纹镂空香囊（清）

这件出土于北京密云董各庄清皇子墓中的玉制香囊，采用镂雕技术，制作极其精美。

Hollow-out Carved Gray Jade Sachet with Lotus Pattern

It was unearthed in Donggezhuang, Miyun County, Beijing, the prince tomb of the Qing Dynasty. This hollow-out carved work of art was extremely exquisite.

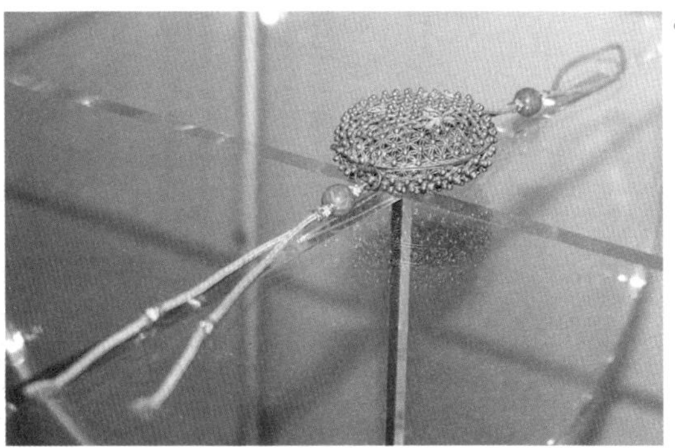

金累丝香囊（清）

清代香囊的种类很多，金质香囊有圆形和长方形两种，一般多镂空，可放入香料或鲜花的花瓣，系于腰间。

Filigree Gold Sachet (Qing Dynasty)

There were many types of sachets in the Qing Dynasty. Hollow-out carved gold sachets were rectangular or round in shape holding petals or perfume, and were tied on the waist.

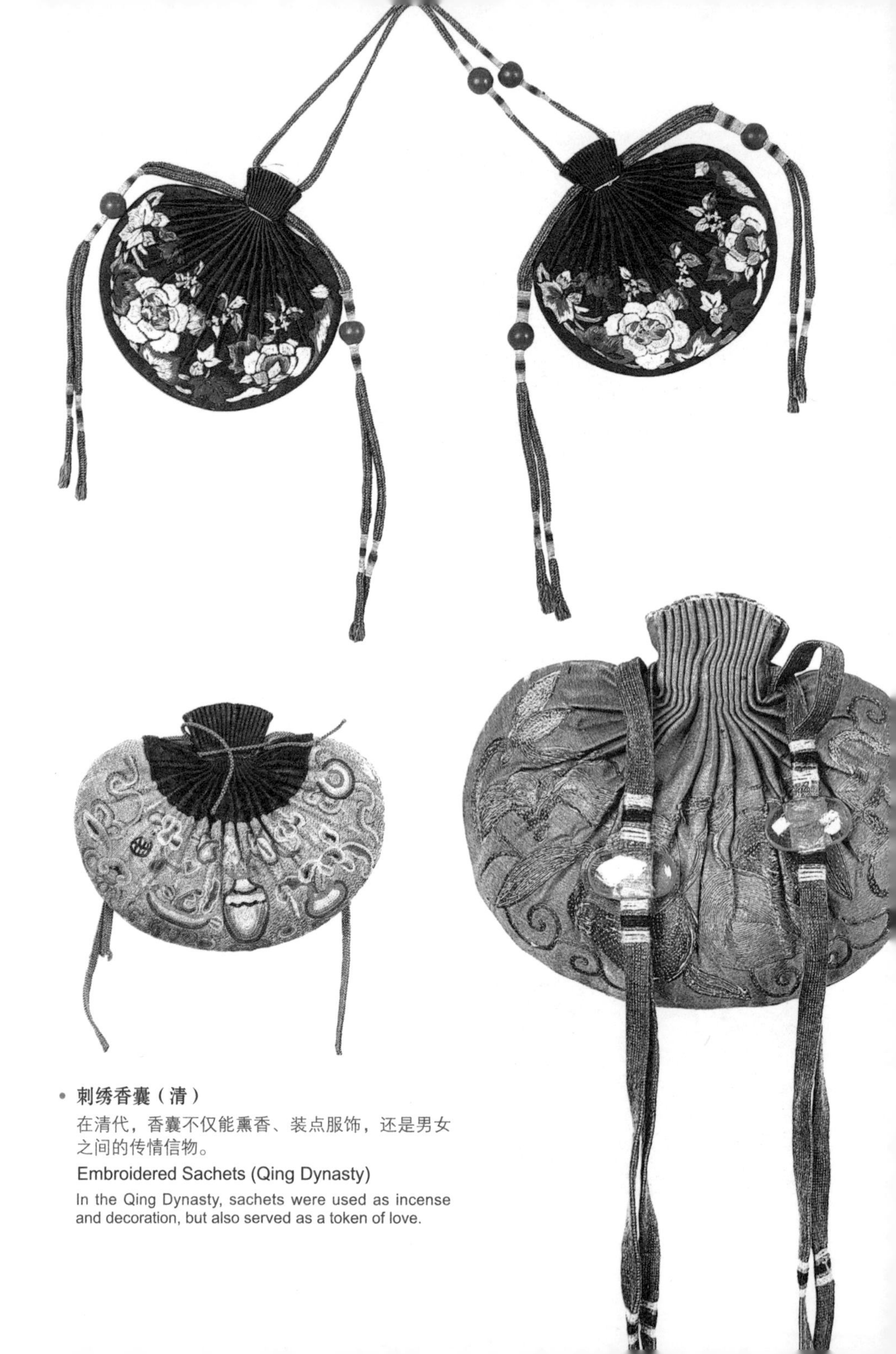

• **刺绣香囊（清）**

在清代，香囊不仅能熏香、装点服饰，还是男女之间的传情信物。

Embroidered Sachets (Qing Dynasty)

In the Qing Dynasty, sachets were used as incense and decoration, but also served as a token of love.

绣荷包

在中国“男耕女织”的封建社会里，刺绣可谓是女人必修的课程，刺绣品遍及生活的方方面面。其中，荷包和香囊都是很重要的绣品。在中国古代，几乎每一个女子都是刺绣的高手，她们从十多岁就开始练习绣花，由母亲、祖母教授。成年时，姑娘们便开始以针线编织着自己的未来，绣荷包、手帕等物品，送给自己的情郎。

Pouch Embroidery

Chinese feudal society has witnessed the traditional custom of "men plowing and women weaving". Almost every woman mastered embroidery in ancient China. Women's embroidered parts were widely used in daily life. Because embroidery has been seen as a compulsory course for women, they started to learn embroidery from their mothers and their grandmothers. When they were in adulthood, they started to embroider handkerchiefs and pouches for their beloved.

- 刺绣绷架
  Embroidery Frame

> 长命锁

长命锁是盛行于明清时期的挂在儿童脖子上的一种吉祥佩饰。人们认为只要将长命锁佩挂在孩子的脖子上，就能平安长大。因此，很多儿童从出生不久（一般是新生儿满百日或周岁举行的仪式中）就挂上了长命锁，一直挂到成年。

长命锁多以银制成，上部为项圈，下部为坠饰物。长命锁的正面多錾刻“长命百岁”“长命富贵”“百家保”“连生贵子”“福胜子仪”等吉祥文字。长命锁的造型很多，有锁形、元宝形、香荷包形、筒形、四方形、长方形、八角形、桃形、蝴蝶形、狮子形、如意形、麒麟送子形等。贵族家的孩子也有用金锁、玉锁的，穷人有用铜锁的，但以用银锁的数量最多。

> Longevity Locks

Longevity lock, an auspicious accessory for children, was popular in the Ming and Qing dynasties. People believed that it helped their children to lead a restful life. Therefore, they wore longevity locks during the neonatal period (generally they would wear them on the occasion of one-hundred-day or one-year-old ceremony) on the neck until they were in adulthood.

Longevity lock was mostly made of silver with a necklet on the upper part of the longevity lock and pendants beneath. Idioms with auspicious meanings were engraved on the front side such as "longevity", "longevity and prosperity" and "getting more sons". Longevity lock had such diverse shapes as gold ingot, lock, pouch, barrel, square, rectangle, octagon, peach, butterfly, lion and *Ruyi*, etc. Children in the noble family wore gold and jade longevity locks while the

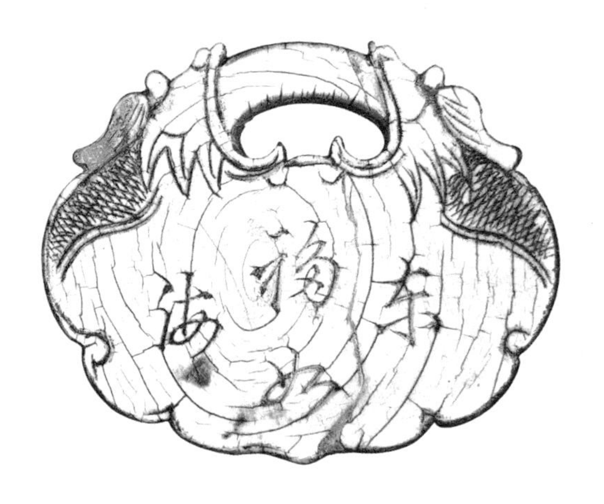

- **福如东海象牙长命锁（清）**

  中国的东海浩瀚无边，因此人们希望福气像东海一样无边无际。

  Ivory Longevity Lock with Design of Happiness as Immense as the East China Sea (Qing Dynasty)

  The East China Sea was so boundless that ancient people metaphorized happiness into the East China Sea.

系挂长命锁的绳索，繁简不一。普通百姓所戴的一般只是一根红色丝带，而富贵人家则用金银打制成链条，也有将珍珠、宝石做成串饰，然后再系于锁上的。

poor wore bronze and silver locks.

The common people often connected the lock with a red silk ribbon while the wealthy wore silver and gold chains. Some people had clusters of pearls and precious stones tied on the lock.

- **银点蓝如意云头长命锁（清）**

  长命锁一般都刻有“长命百岁”四个字，即长寿到百岁。该银锁的做工十分讲究，银链上挂有六个红玛瑙坠，用两个小银元做挂钩。锁的正面点蓝打底，上方是吉祥云图，下方是蝙蝠，两旁是吉祥花卉，中间写着“长命百岁”四个字。银锁的外形为如意云头，做工十分精致。银锁下坠有花篮、鱼、铃铛等挂件，丰富了锁的造型。

  Silver Stippling Blue Longevity Lock with *Ruyi* Head (Qing Dynasty)

  Most of longevity locks were engraved with four Chinese characters "*Changming Baisui*" meaning "longevity". This silver lock was made exquisitely. Six red agates pendants were hung on the silver chain hooked by two small silver coins. The lock was stippled blue with four Chinese characters in the center, auspicious cloud patterns above, floral patterns on both sides and bat patterns below. The outline of the lock was a head of *Ruyi* in shape (an ornamental object, a symbol of good luck).Such accessories as fish, basket and bell were hung below the lock which diversified the design of the longevity lock.

• **蝴蝶纹银长命锁（清）**

蝴蝶在中国古代被视为美好的象征，常用以比喻美好的爱情与姻缘。同时，蝴蝶的“蝶”又与“耋”同音，中国古代称七八十岁的年纪为“耋”，因此蝴蝶又有长寿之寓意。

Sliver Longevity Lock with Butterfly Pattern (Qing Dynasty)

The butterfly has been seen as an emblem of sweet love and happy marriage. Because the character of "butterfly" (*Die*) is the homonym of "To or 80-year-old age" in Chin`ese, the butterfly also means longevity.

• **麒麟送子银长命锁（清）**

麒麟送子的造型在长命锁中最为常见。

Silver Longevity Lock Engraved with Pattern of Kylin Offering a Son (Qing Dynasty)

It was the most common design on longevity lock.

- 连中三元银长命锁（清）

中国古代的国家级的科举考试有三个等级。乡试第一称“解元”，会试第一称“会元”，殿试第一称“状元”。若乡试、会试、殿试皆为第一名，则称“连中三元”。

Silver Longevity Lock with the Pattern of "Ranking the First in Three Examinations" (Qing Dynasty)

In ancient Chinese imperial examination system, there were three levels of examinations. The low level was the county examination. The intermediate level was the provincial examination. The top level was the imperial examination. The one who ranked the first in the three levels of examinations, it was called "ranking the first in three examinations".

- 麒麟送子银长命锁（清）

此银锁以錾花工艺成型，造型古朴，并用玛瑙、玉片等加以装饰，加以多种吉祥坠饰，十分珍贵。

Silver Longevity Lock Engraved with Pattern of Kylin Offering a Son (Qing Dynasty)

This longevity lock was chiseled and decorated with agates, jade plates and a variety of auspicious pendants. Its shape was simple and unsophisticated.

- **福在眼前银长命锁（清）**

这件银锁的图案由蝙蝠、桃、钱组成。“蝙蝠”的“蝠”与“福”同音；“桃”有“长寿”之意；“钱”古时又称“泉”，与“全”同音。蝙蝠、桃、钱而又各自成双，因此寓意“福在眼前”或“福寿双全”。此锁形象生动，做工精巧，尤其是下面坠的五个圆环状铜钱，与上面的蝙蝠搭配起来，十分雅致。

Silver Longevity Lock with the Combined Pattern of Bat, Peach and Money (Qing Dynasty)

In Chinese, the character of bat (*Fu*) is a homonym of happiness (*Fu*). The peach has the meaning of longevity. In ancient China, the pronunciation of "money" was similar to "well" and "full" (*Quan*). The design of the lock was delicate and vivid. Five money-shaped pendants matching with the bat looked elegant. This pattern expressed the meaning of "happiness and longevity".

- **金鱼进宝银长命锁（清）**

因鱼产子较多，所以中国古代以其象征多子，同时也作为丰收、富裕的象征。中国很多吉祥图案和饰品都离不开鱼，比如“连年有余”“鲤鱼跳龙门”等。

Silver Longevity Lock with Fish Offering Treasures Pattern (Qing Dynasty)

The fish spawning symbolized more offspring, good harvest and wealth.Chinese auspicious designs were inseparable from the fish pattern. Fish pattern was used to express the meaning of surplus year after year. Also, there was an auspicious pattern named carp jumps over the dragen gate (meaning getting promotion).

长命缕

长命锁虽在明清时才开始盛行，但它的历史十分悠久，其渊源可以追溯到汉代的“长命缕”。当时，由于战争频繁，自然灾害、瘟疫不断，百姓们渴望平安，便用五色彩丝编成绳索，缠绕在妇女和儿童的手臂上。这种彩色的丝绳，当时称之为“长命缕”或“长生缕”。到了明代，这种风俗逐渐改变。成年男女极少佩戴长命缕，其逐渐成为一种未成年人的颈饰，也就是后来的长命锁。

Longevity String

Although longevity lock prevailed in the Ming and Qing dynasties, the origin of longevity lock can be traced back to the Han Dynasty during which it was called "longevity string". Women and children wore the hand-woven five-colored string around the wrists, known as "longevity string". This custom gradually evolved in the Ming Dynasty when children still wore it while adult men and women changed to wear the accessory around the neck which was the prototype of longevity lock.

帽花

帽花是指装饰在帽子上、额间的饰品，其作用跟“长命锁”十分相近，多是在小孩出生过满月、百天、周岁等喜庆节日时作为吉祥物。帽花一般以神仙、花卉、动物、八宝等吉祥物为主，还有的帽花由“长命百岁”“竹报平安”“状元及第”“荣华富贵”等吉祥词语组成。民间百姓认为给小孩戴上这些寓意吉祥的帽花，就可以保佑其吉祥平安。除了小孩，妇女也使用帽花，其图案一般以团花、团寿、团鹤、福禄寿喜等吉祥物为主，此外还有牡丹、菊花、蜘蛛、蝴蝶等动植物。戴帽花的女子一般年龄较大。很多帽花是作为儿媳在婆婆过寿时赠送的寿礼，能起到装饰的作用。

- **吉庆有余纹银帽花（清）**

中国汉字中，“鱼”与“余”谐音，所以鱼有丰收、富余之寓意。

Silver Badges with Fish Pattern (Qing Dynasty)

In Chinese, the characters of "fish" and "surplus" were homonym, so fish was used to symbolize surplus and harvest.

- **白玉帽花（清）**

白玉帽花主要装饰在女子的帽子上。

White Jade Badges (Qing Dynasty)

They were mainly decorated on women's hats.

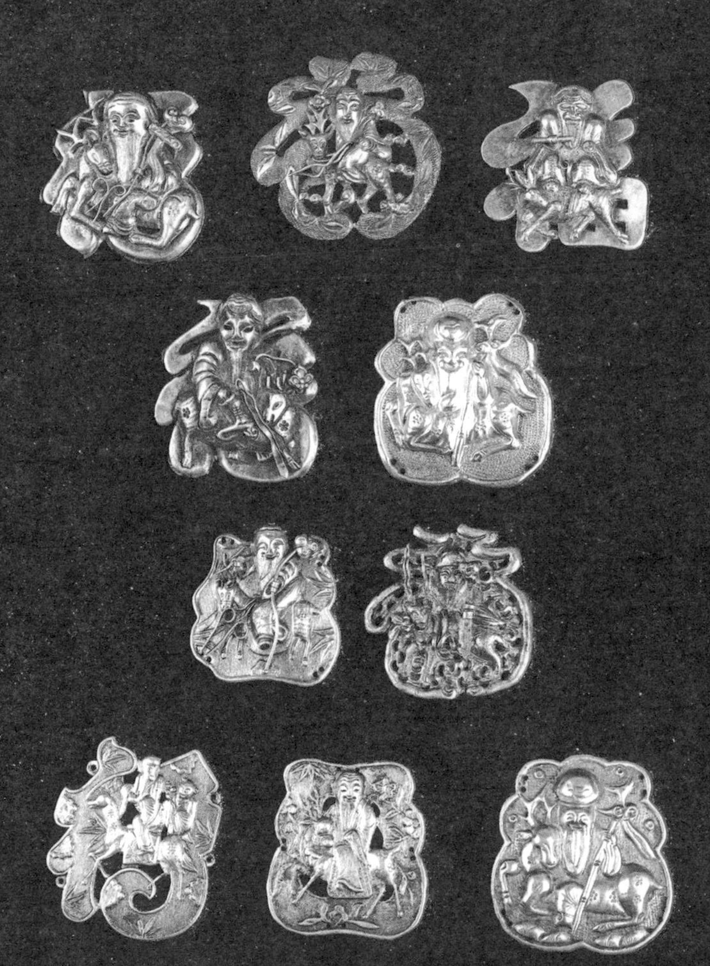

老寿星纹福字银帽花（清）

这组老寿星纹福字帽花，将文字巧妙经营加以细微装饰，反映了中国百姓追求平安幸福的美好愿望。

Silver Badges with the God of Longevity Pattern (Qing Dynasty)

Ingenious Chinese characters and subtle ornamentation could be found on the patterns. It reflected the aspiration of Chinese people in the pursuit of peaceful and happy life.

Badge Decoration on the Hat

The badge was used to decorate the hat. It played a similar role as longevity lock, that is, it was an auspicious accessory for children. When celebrating important days such as the full month, one hundred days, one year old and festivals, children would wear badges. Badges could be fallen into two types. One was decorated with auspicious patterns such as gods, Buddha statue, flowers, animals and eight treasures. The other was ornamented with auspicious words to express the meaning of longevity, prosperity and safety. Common people believed that children would live safe and sound as long as wearing these auspicious badges. Elder women also wore badges. They were used as birthday presents for mother-in-law when her daughter-in-law offered birthday congratulations. Auspicious Patterns included flowers, cranes, peonies, chrysanthemum, spiders, butterflies, animals and plants.

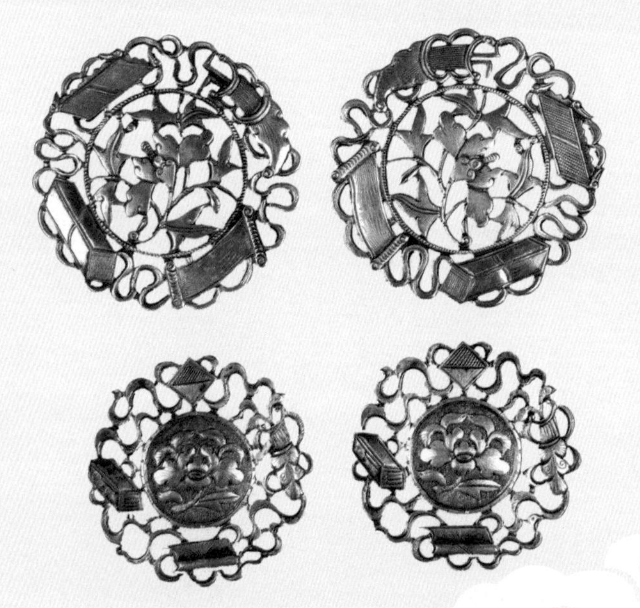

• 琴棋书画银帽花（清）

Silver Badges with Lute-playing, Chess, Calligraphy and Painting Pattern(Qing Dynasty)

> 鼻烟壶

鼻烟是将优质的烟草研磨成的极细的粉末，其中加入了麝香等名贵药材，可以用鼻嗅服。17世纪，鼻烟在欧洲开始流行，明末清初传入中国后，装鼻烟的盒子渐渐

> Snuff Bottles

Snuff is a kind of very fine powder ground by high-quality tobacco, to which some precious medicinal materials like musk are added. In 17th century, snuff became popular in Europe. Snuff bottle has been orientalized step by step since snuff was introduced into China in the late Ming and early Qing dynasties. Snuff bottle was not only used as a container for snuff storage but also used to show wearers' social status. Emperor Qianlong of the Qing Dynasty (1711-1799) often awarded dukes and ministers the snuff bottle. Snuff bottle was small and flat

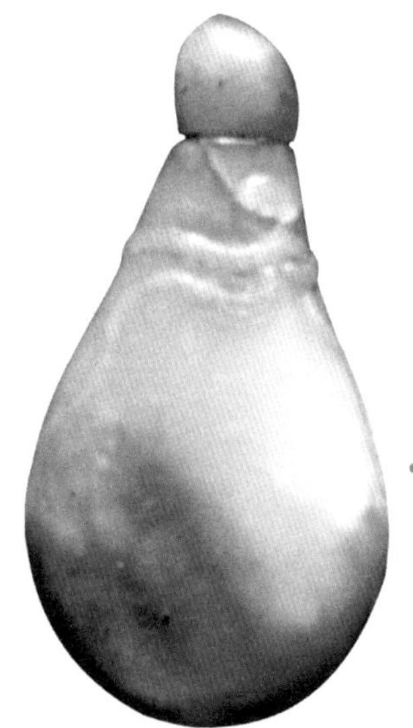

• **白玉随形鼻烟壶（清）**

此鼻烟壶通体曲线流畅，形制简朴，壶身上敛下阔，十分精巧。

White Jade Snuff Bottle (Qing Dynasty)

The curve of the snuff bottle was smooth. This snuff bottle was narrow on the upper part and wider on the lower part. The design was simple but delicate.

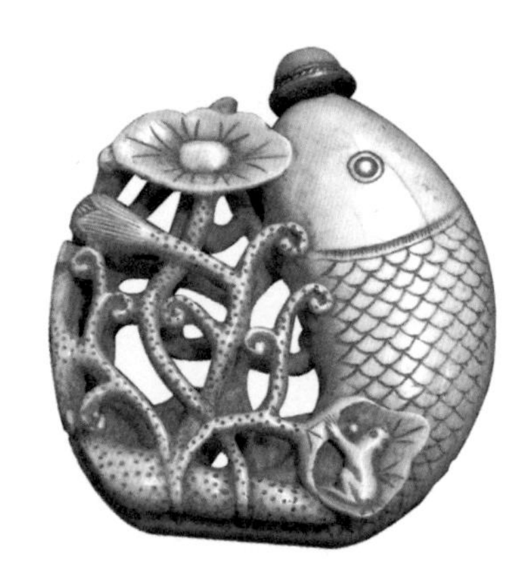

• **象牙鼻烟壶（清）**
象牙器最初只为皇族所特有，后来逐渐为民间所青睐，但由于象牙材料的稀缺，这种器具仅在富贵人家才可见到。
Ivory Snuff Bottle (Qing Dynasty)
Ivory items were initially for the royal's exclusive use. Later, they were favored by common people. Because of the scarcity of the ivory, only the rich use it.

东方化。鼻烟壶不仅是可以用来储存鼻烟的器具，也可以用于玩赏和显示身份地位，乾隆皇帝就常以鼻烟壶赐赏王公大臣。为了便于携带，鼻烟壶的形状通常是小而扁的，一般用绳挂在身上或别在腰间。清代道光年间由宫内养心殿造办处特制鼻烟壶，壶盖内附小细匙，以便舀取烟粉。康熙年间，清宫造办处制造的玻璃、景泰蓝等各种材质的鼻烟壶，开辟了工艺美术的一项新门类。

鼻烟壶的材质主要有瓷、玻璃、玉、玛瑙、木、竹、象牙、

in shape tying on the waist. During the reign of Emperor Daoguang of Qing Dynasty, a special kind of snuff bottle with a tiny spoon in it to scoop the tobacco powder was made by the Hall of Mental Cultivation. The manufacturing department of the court made the snuff bottle in a variety of materials such as glass, cloisonné and so on, which opened a new page in arts and crafts.

Snuff bottle was made of a variety of materials such as porcelain, glass, jade, agate, wood, bamboo, ivory and cloisonné. Different shapes could be found such as square, round, figures, gourd-shaped and all kinds of fruit

景泰蓝等。它的造型非常丰富，有方形、葫芦形、圆形、瓜果形、人物形等，还有各种仿古造型。鼻烟壶的琢磨工艺精湛，有的随形而做，有的顶盖镶嵌宝石、碧玺、珊瑚等。

鼻烟壶是融合了中国雕刻、书法、绘画、烧瓷、镶嵌等传统技艺的工艺品。如今，人们嗜用鼻烟的习惯几近绝迹，但鼻烟壶却作为一种精美的艺术品流传下来，被誉为"集各国多种工艺之大成的袖珍艺术品"。

shapes. Some of the bottles were made at random. There was a variety of antique shapes. Some of the lids were inlaid with precious stones, tourmaline and corals. In general, the carving and polishing techniques were very exquisite.

Snuff bottle is a fusion of the traditional Chinese sculpture, calligraphy, painting, porcelain, mosaic and other arts and crafts. Today nobody is addicted to snuff any more but the snuff bottle as a beautiful work of art has been handed down, known as "a tiny work of art combined a lot of techniques from all over the world".

• 红玻璃菱纹鼻烟壶（清）
Red Glass Snuff Bottle with Diamond Pattern (Qing Dynasty)

• 青花瓷花鸟人物纹鼻烟壶（清）
Blue-and-white Porcelain Snuff Bottle with Flower, Bird and Figure Patterns (Qing Dynasty)

• **玻璃胎画珐琅梅花纹鼻烟壶（清）**

清代的玻璃器品种非常丰富，每年向外地输出数千件。其工艺精美，装饰多采用雕刻、描彩、泥金和珐琅彩等方法。

Glass Snuff Bottle with Enamel Plum Pattern (Qing Dynasty)

A variety of glass wares appeared in the Qing Dynasty. Several thousand glass wares were transported to every corner of China every year. The decorative techniques were exquisite including colored painting, carving, enamel and powered gold painting.

• **粉彩描金花卉鼻烟壶（清）**

粉彩即以粉彩为主要装饰手法的瓷器品种，给人以秀丽雅致、粉润柔和之感。

Famille Rose Snuff Bottle with Floral Pattern (Qing Dynasty)

Famille rose is one of the porcelain decorative techniques, which gives a sense of beauty, elegance and gentleness.

• **青花瓷兽耳鼻烟壶（清）**

青花瓷是中国瓷器的主流品种之一，烧成后釉面清爽透亮，纹饰灵动而不失规矩，表面的青花发色含蓄沉静，呈色稳定。

Blue-and-white Porcelain Snuff Bottle with Bulgy Beast Ear on Each Side (Qing Dynasty)

Chinese blue and white porcelain has been one of the mainstreams. After being burnt, the glaze is translucent. The pattern looks vivid but in good order. The color of finish appears subtle, quiet and enduring.

内画鼻烟壶

内画鼻烟壶出现于清代末期，是用特制的微小勾形画笔，在透明的壶内绘制而成的。内画鼻烟壶的内壁最初较为光滑，不易附着墨和上色，因此只能画一些简单的画面和图案，如龙、凤和简笔的山水、人物等。后来，艺人们掌握了新的技术，使鼻烟壶的内壁呈乳白色的磨砂玻璃，细腻而不光滑，极易附着墨色。从此，内画鼻烟壶便出现了一些比较精细的作品，成为诗、书、画并茂的艺术精品。

Inner Painting of Snuff Bottle

The inner painted snuff bottle appeared in the late Qing Dynasty. It was painted on the transparent inner side of the bottle. Initially, the inner wall of the bottle was too smooth to put color on; therefore, only simple drawings were painted such as dragon, phoenix, simplified landscapes and figures. Later craftsmen mastered the new technique. After the inner wall of the bottle was replaced by the milky white ground glass with fine but unsmooth texture, ink was adherent to the inner wall easily. Snuff bottle became the fine work of arts which integrated poetry, calligraphy and painting.

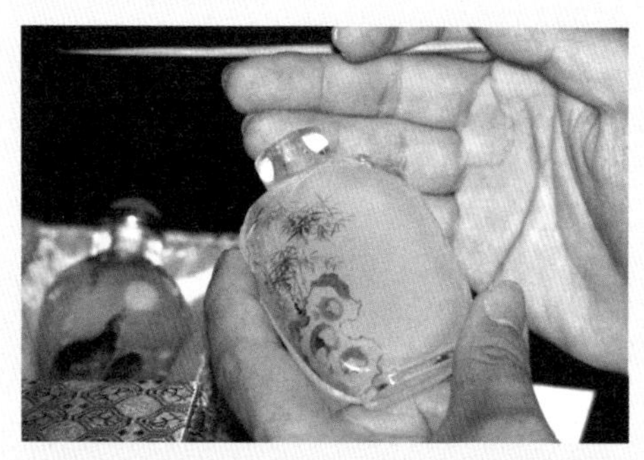

- 鼻烟壶艺人正在进行内壁绘画
The Craftsman is Painting on the Inside of the Snuff Bottle

- 内画鼻烟壶
Inner Painted Snuff Bottle

> 压胜钱

压胜钱并非流通货币，而是中国古代民间流行的一种吉祥佩件，挂在身上用以祝福祈愿。压胜钱以纹饰和造型来求福、求寿、求财、求子、求平安等，表达了百姓们的美好愿望和对生活的追求。

压胜钱起源于西汉，当时只是在铸造钱币时在钱币背面加铸日、月、星、龟蛇、宝剑等图案。唐宋时期的压胜钱以镂空雕刻为主，纹样主要包括花草、虫、鱼、蝶、龙、凤、狮、鹿、人物故事等。这类压胜钱主要起装饰作用，同时寓意吉祥、太平。

压胜钱盛行于明清时期，这一时期的压胜钱数量最多，题材也最丰富，所用的范围也越来越广，诸如开炉、镇库、馈赠、赏赐、祝

> Auspicious Coins

Auspicious coins which were not for circulation were popular among common people. They were worn on the body to pray for happiness. Depending on different patterns and shapes of auspicious coins, common people prayed for happiness, longevity, fortune, expecting sons and safety, which expressed their pursuit of happy life.

Auspicious coins were originated in the Western Han Dynasty when some patterns such as the sun, the moon, the star, tortoise and sword were casted on the back of the coins. Most of auspicious coins in the Tang and Song dynasties were made by openwork carving. Patterns included flowers, insects, fish, butterflies, dragon, phoenix, lion, deer and figures in the stories. In addition to the auspicious meaning, coins at that time were used for decoration.

- **常平五铢钱（南北朝）**

常平五铢钱原是借用南北朝时期囤积粮食的仓库“常平仓”之名。后来人们认为其名称有“平安常在”之意，因此常常将其携带在身边作为护身符使用，寓意吉祥平安。

Changping Wuzhu Coin (Northern and Southern Dynasties)

Changping was the name of a grain depot in the Northern and Southern dynasties. One believed that the name enjoyed the good meaning of safety hence it was used as an amulet.

福、玩赏等，都铸压胜钱，其形制、铭文、纹饰都达到了精美绝伦的地步。此时期压胜钱的材质主要为黄铜，少数以金、银、铁、铅制造，还有的填珐琅彩，铸造技术极为精湛。清康熙以后，压胜钱的图案更为多样，内涵也更为丰富，除以前历代所用图案外，人们又将宝物、书画、植物、水果、乐器等铸于钱上。

Auspicious coins prevailed in the Ming and Qing dynasties, the number of which increased dramatically. They were used in a wide range such as putting in the warehouse, giving as gifts, and praying for happiness. They were made of brass, gold, silver, iron, lead as well as enamel. The exquisite casting technology made the patterns delicate. After Emperor Kangxi of the Qing Dynasty, patterns became rich in connotation. Additional patterns such as treasures, calligraphy, plants, fruits and musical instrument enriched patterns of previous dynasties.

- **福禄黄铜压胜钱（清）**

铜钱的上方为“福”字，下方用一只鹿代替了“禄”。中国古代尊鹿为瑞兽，有长寿之意；同时“鹿”与“禄”同音，又有祝福仕途顺利之寓意。

Auspicious Coin with the pattern of one character "happiness" and a deer (Qing Dynasty)

Taking advantage of homonyms, the deer pattern referred to "official's salary in feudal China". The deer was seen as an auspicious animal in ancient Chinese mind. It symbolized longevity and promising career.

- **一品当朝状元及第黄铜压胜钱（清）**

“一品鸟”是鹤的雅名，“一品”意为地位最高。在清朝的官服上，文官的一品官服绣的就是鹤。“当朝”有执掌国政之意。

Brass Auspicious Coin Symbolizing Wielding National Power (Qing Dynasty)

The crane pattern was the embroidered on the first rank civil official's costume in the Qing Dynasty. The first rank was the highest position in the court representing "wielding national power".

- **积玉堆金黄铜压胜钱（清）**

“积玉堆金”一词是比喻大富大贵，财源茂盛，钱多得堆起来。

Auspicious Coin with Characters of "Piles of Jade and Gold" (Qing Dynasty)

It represented wealth and prosperity.

- **虎镇五毒黄铜压胜钱（清）**

“五毒”是指蝎子、蜈蚣、蛇、蜥蜴、蟾蜍这五种动物。在中国古代，每年的农历五月初五这一天将“五毒”的纹样做成肚兜、荷包或压胜钱等佩带在儿童身上，作为护身符，寓意保孩子平安。

Brass Auspicious Coin with Five Poisonous Creatures Design (Qing Dynasty)

Five poisonous creatures refer to scorpion, viper, centipede, house lizard and toad. Ancient people believed that poison would be immense on the day of May 5th in Chinese lunar calendar. On this day, people had their children wear "*Duduo*" (a kind of underwear), pouches or auspicious coins with the pattern of five poisonous creatures as amulets.

- **十二生肖黄铜压胜钱（清）**

中国古人用鼠、牛、虎、兔、龙、蛇、马、羊、猴、鸡、狗、猪等十二种动物排列组合成十二生肖（即属相）。十二生肖压胜钱在明清时期十分普遍，寓意吉祥平安。

Brass Auspicious Coin with Chinese Zodiac Pattern (Qing Dynasty)

The Chinese zodiac includes twelve animals. They are the rat, ox, tiger, rabbit, dragon, snake, horse, sheep, monkey, rooster, dog and pig. Zodiac patterns were popular in the Ming and Qing dynasties symbolizing safety.

- **珐琅彩银压胜钱（清）**

珐琅彩是将铜胎画珐琅技法移植到瓷胎上的一种工艺。珐琅彩瓷质细润，彩料凝重，色泽鲜艳亮丽，画工效果精致，是专供清代宫廷御用的一种精细彩绘瓷器。部分产品也用于犒赏功臣。

Enameled Silver Auspicious Coin (Qing Dynasty)

The technique of the enameled copper was used on the porcelain. The enameled porcelain which had beautiful and bright color, fine finish and exquisite painting skill was exclusively for the emperor of the Qing Dynasty. Some of them were rewarded meritorious officials.

- **八卦纹银压胜钱（清）**

八卦纹属中国道教文化范畴，中间由黑、白两个鱼形纹组成，寓意阴阳轮转，相辅相成，体现万物生长变化的哲理。此外，八卦纹也是古代民间镇邪的吉祥符号。

Silver Auspicious Coin with Eight Diagrams Design(Qing Dynasty)

Eight Diagrams (eight combinations of three whole or broken lines formerly used in divination) is a part of Chinese Taoist culture composed of a white and a black fish-like pattern, symbolizing *Yin* and *Yang* rotation to be supplementary to each other, which is the philosophy related to the growth and changing of all things. It is also the auspicious symbol of guarding the house.

> 腰挂

腰挂是悬挂在腰带上的实用物品及装饰物。腰挂的种类繁多，如佩刀、火镰、针筒以及各种装饰品等。

腰挂最初是人们为了生活方便而随身携带的日常用品。男子多悬挂小型武器及点火工具，女子则佩带针筒之类的生活用具。后来，这些腰挂逐渐演变成装饰物、吉祥物、护身符。

> Accessories on the Waist

There was a variety of accessories and daily necessities hung on the waist such as walking sabre, steel for flint, needle holder and other accessories.

Initially people wore daily necessaries on the waist for the sake of convenience. Men wore small-sized weapons and fire making tools while women had needle holders with them. They gradually evolved into accessories such as mascots and amulets.

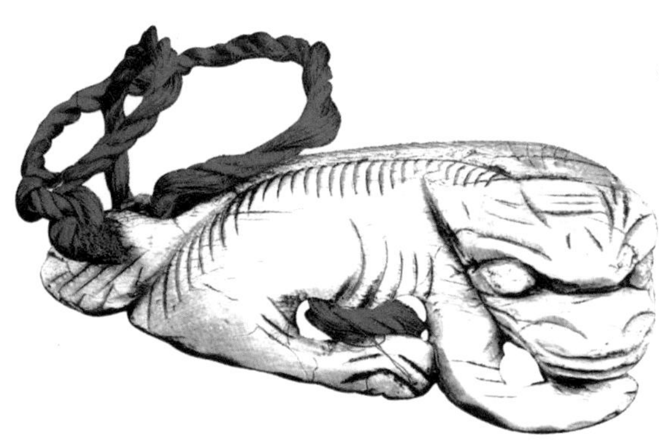

• **虎形象牙腰饰（清）**

老虎被称为“兽中之王”，被认为可以镇宅辟邪，保佑安宁。为此，父母常会给小孩穿戴虎头鞋、虎头帽。

Ivory Waist Accessory in Tiger Shape (Qing Dynasty)

The tiger is known as the "king of the beast". It's commonly believed that the tiger can guard the house to bless people with safety. Children were often dressed in shoes and hats with tiger design.

人物形银针筒（清）

针筒有各种造型和图案，且几乎每个针筒的图案都寓意吉祥。如童子手捧莲花或笙，寓意连生贵子；童子脚下踩着元宝，寓意“送财童子”。

Figures Shaped Silver Needle Holders(Qing Dynasty)

Diverse as the patterns and shapes of needle holders were, each pattern had the auspicious meaning such as a boy offering lotus meaning having more sons and shoe-shaped silver ingot under a boy's feet referred to as "a boy sending a fortune".

• 银连生娃（清）

连生娃又称“喜相逢”，即两个或多个笑容满面的童子，寓意连生贵子。

Silver Accessory with Design of a String Boys (Qing Dynasty)

More than two boys with smiling faces were linked together symbolizing more sons to be born.

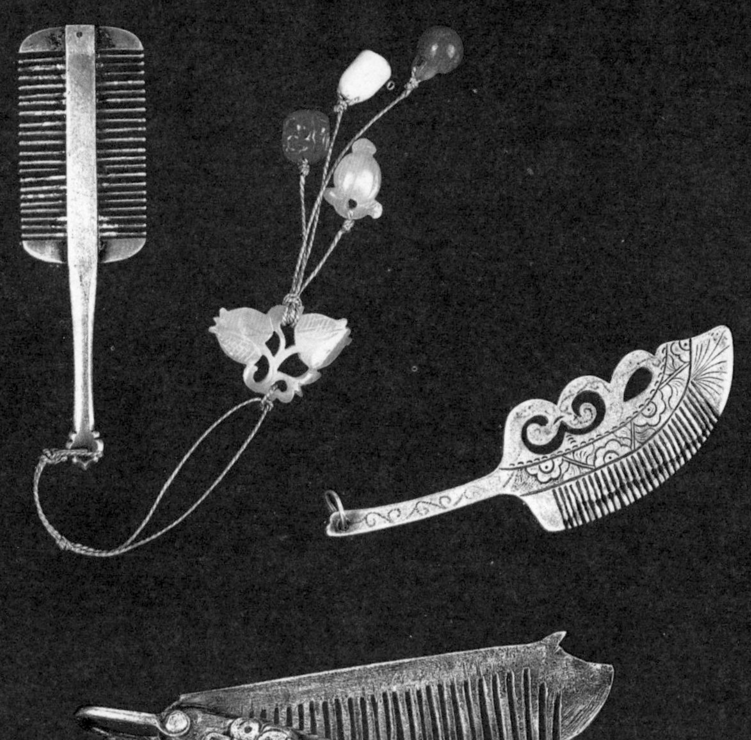

银梳（清）

古代女子常将梳子挂在腰间，最初是为了使用方便，后来也可用于装饰，同时寓意吉祥。梳子是由上往下梳，有疏通之意。还有一种豆荚造型的梳子，寓意四季通顺。

Silver Comb (Qing Dynasty)

Ancient women often hang their combs on the waist for the sake of convenience but later they were used for decoration with auspicious meanings. Combing from top to bottom symbolized going on well. There was a bean-shaped foldable comb meaning going on well in four seasons.

鱼纹银腰饰（清）

此腰饰寓意多子多孙。鱼下垂挂两个花生，代表连生；花生下垂挂两个石榴，代表多子。

Waist Accessory with Fish Design (Qing Dynasty)

This waist accessory symbolized more descendants. There were two peanuts hanging below fish and two pomegranates at the bottom representing more sons.

银火镰（清）

火镰是古代的取火器物，由于打造时把形状做成镰刀的形状，与火石撞击时能产生火星而得名。

Silver *Huolian* (Flint) (Qing Dynasty)

Huolian was a tool for making fire, in sickle shape, hence its name.

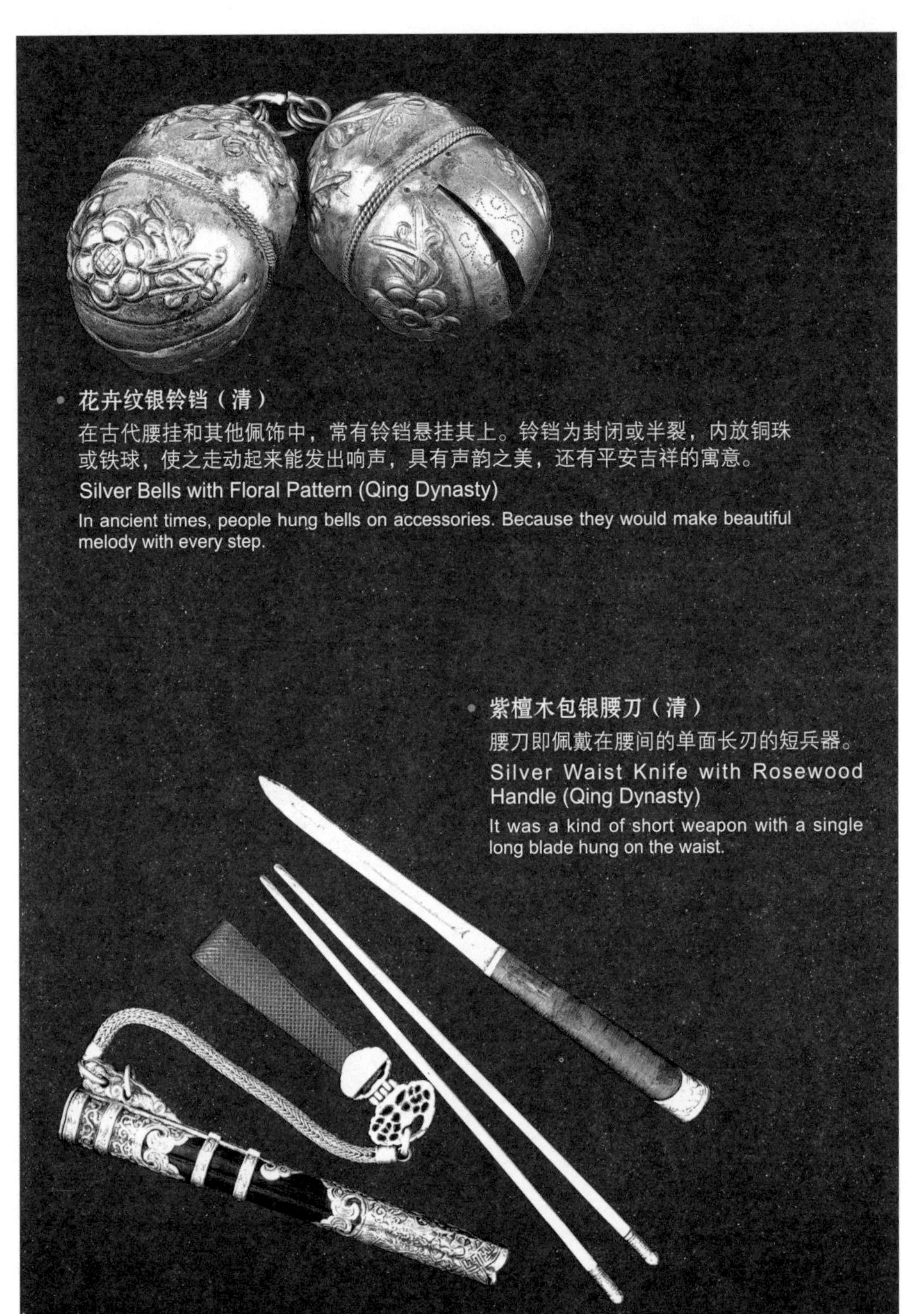

花卉纹银铃铛（清）

在古代腰挂和其他佩饰中，常有铃铛悬挂其上。铃铛为封闭或半裂，内放铜珠或铁球，使之走动起来能发出响声，具有声韵之美，还有平安吉祥的寓意。

Silver Bells with Floral Pattern (Qing Dynasty)

In ancient times, people hung bells on accessories. Because they would make beautiful melody with every step.

紫檀木包银腰刀（清）

腰刀即佩戴在腰间的单面长刃的短兵器。

Silver Waist Knife with Rosewood Handle (Qing Dynasty)

It was a kind of short weapon with a single long blade hung on the waist.

> 发饰

中国古人除了创制出了千姿百态的发式外，还在美丽的发式上插戴各种饰品，如笄、簪、钗、步摇等，既能固定发髻，又能提升美感。

笄

笄是中国最古老的一种发簪，用来插住挽起的长发，或插在帽子上。商周时期的男女都用笄，女子用来固定发髻，男子用来固定冠帽。古时的冠很小，必须用双笄从左右两侧插进发髻加以固定，故称为“衡笄”。

笄除了具有固定和装饰发髻的作用，还是“分贵贱，明等威”的工具。不同身份的人所用的笄的材质也有所不同，皇帝、皇后和一些

> Hair Ornaments

In addition to various hairdos, ancient people created beautiful and diverse hairpins to fasten hair buns.

Ji (Hairpin for the Bun and Crown)

Ji, the oldest hairpin in China, was used to make hairdo and pin on the hat. Both men and women used *Ji* in the Shang and Zhou dynasties. The ancient crown was so small that two hairpins had to be inserted into the bun from left and right sides to fasten the crown.

The emperor, empress and royal family members wore jade *Ji*. Scholar-bureaucrat's wife wore ivory *Ji* while the common people used bone *Ji*. Therefore, *Ji* acted as a tool to distinguish people's social status along with its basic role of decoration and hairdo.

The character *Ji* originally appeared in the Shang and Zhou dynasties but

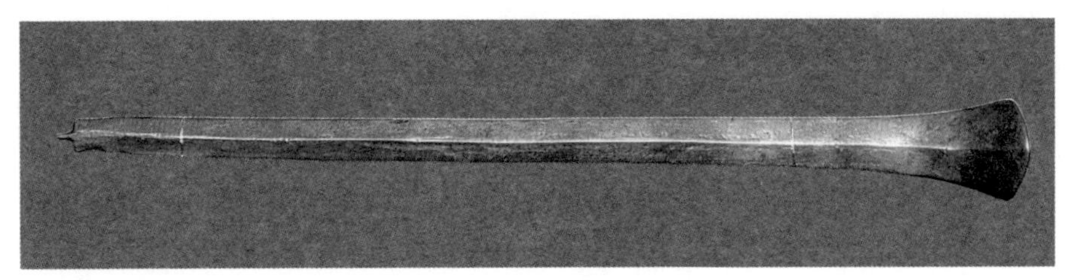

- **金笄（商）**

此笄两面不一，一面光滑，一面有脊，截断面呈纯三角形，头部宽，尾部逐渐变窄，便于穿插。其尾端有一长约0.4厘米的榫状结构，用于镶嵌其他的装饰品。

Gold *Ji* (Shang Dynasty)

This hairpin was different on both sides. One was smooth, the other had a ridge. The cross section was triangular in shape with wide head and narrow end. It was designed to wear easily. There was a tenon-shaped structure about 0.4cm in length for setting other ornaments.

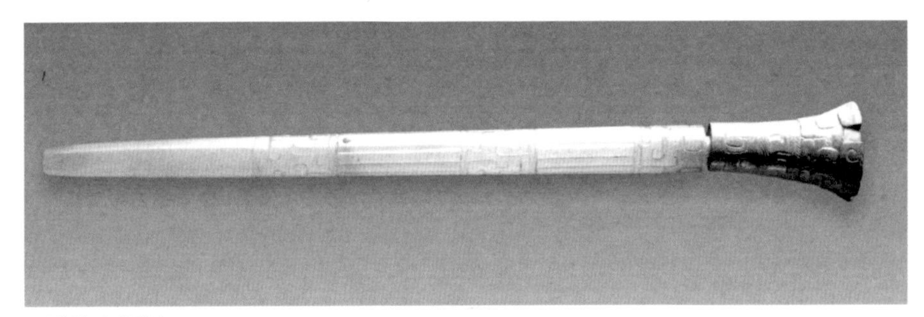

- **玉笄（春秋）**

Jade *Ji* (Spring and Autumn Period)

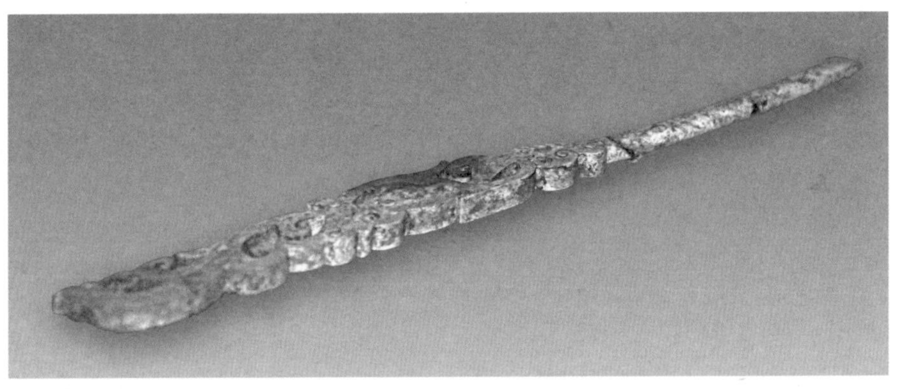

- **玉笄（西汉）**

Jade *Ji* (Western Han Dynasty)

有身份的皇亲国戚用玉笄，士大夫的妻子用象牙笄，平民女子只能用骨笄。

“笄”是商周时期的称谓，战国以后多称为“簪”，男女皆可使用，男子用于拴冠，女子则用于固髻，一般以材质区分身份等级。最初的发笄多用兽骨制成，也有玉笄、竹木笄、金银笄、象牙笄、玳瑁笄等。笄的形状有凤形、蛙形等，上端大都雕刻凤鸟、鸳鸯等动物图案，以及几何图案。

it was replaced by a synonym after the Warring States Period. Men wore *Ji* on the crown while women pinned *Ji* on buns. The social status could be distinguished by its material. *Ji* was made of bones, jade, bamboo, silver, gold, ivory or tortoiseshell. The phoenix-shaped and frog-shaped *Ji* was engraved phoenix, mandarin ducks, other animals and geometrical patterns on the upper part.

• **凤首玉笄（唐）**

唐代的玉笄一般用极薄的白玉制成，形式多样，有半月形、叠叶形及其他样式。

Jade *Ji* with Phoenix Heads (Tang Dynasty)

Jade hairpin in the Tang Dynasty was generally made of a very thin white jade with a variety of shapes such as the half-moon and the overlapping leaf.

笄礼

古代的孩童是不梳发髻的，将头发扎成小丫角，显得活泼可爱。直到笄礼举行后，不论男女，都要梳起发髻来。笄礼即成年礼，规定女子年满15岁、男子年满20岁时即要举行，梳绾发髻，以笄固定。笄礼的举行，标志着从此开始接受社会规范，男子将成为温恭贤良的君子，女子可以许嫁。

未许嫁的女子的笄礼，仅仅是请人象征性地给她梳个发髻，插上发笄，礼毕后仍要取笄解髻，恢复原来的发式。已许嫁的女子的笄礼，隆重而喜庆，除了插上发笄，还要在发髻上束一根彩色的缨线，作为“名花有主”的标志。凡看到此种标志，其他男人不能问津打扰；女子要深居闺室，直到洞房花烛夜才能由新郎亲自解下束发之缨。

Ceremony for Wearing the Hairpin

Children in ancient times wore braids instead of buns. Until ceremony for wearing the hairpin was held, men were at the age of 20 and women 15, they would wear buns symbolizing reaching adulthood. Women at that age could get married.

The ceremony for unmarriageable women was only a symbolic one. Their hair bun and hairpin had to be removed after the ceremony ended. The ceremony for marriageable women was festive. Besides the hairpin, a colored tassel was tied on their bun to show others that they are ready to be married. It acted as a symbol to keep other men away. Women would not leave their bedrooms until their husbands came to untie the tassel on the night of wedding ceremony.

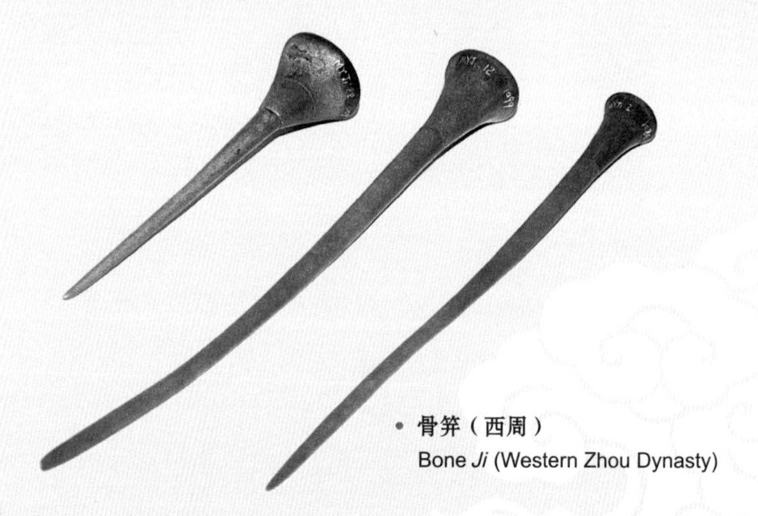

• 骨笄（西周）
Bone *Ji* (Western Zhou Dynasty)

簪

Zan (Hairpin)

簪是中国古代男女最常用的首饰之一，其前身是发笄，战国以后开始称“簪”。

簪的最初用途仅仅是绾束头发，进入阶级社会以后，则逐渐演变成炫耀财富、昭明身份的一种标志。上古时期的石笄、蚌笄、竹笄、木笄、骨笄等相继被淘汰后，取而代之的是玉簪、金簪、银簪、翠羽簪、玳瑁簪或金镶宝石簪等。在各类的发簪中，玉簪和金簪均是非常昂贵的饰品，普通人家很少用得起，而银簪多为中下层妇女使用。

簪的式样十分丰富，其主要变化大都集中在簪首。簪首有各种各样的形状，如花鸟鱼虫、飞禽走兽等，其中以凤簪为最多，制作也最为精致。

Zan was one of the most common hair accessories for men and women in ancient times. Before Warring States period, hairpin was referred to as *Ji*. Its name was replaced by *Zan* after the Warring States Period.

Zan was initially used as making hairdo. *Zan* became an identification showing off the wear's wealth and social status in class society. *Zan* which was made of jade, silver, gold, kingfisher feathers and tortoiseshell gradually replaced the materials of stone, mussel, bamboo, wood and bone. Jade and gold *Zan* were the most precious. They were not affordable by ordinary people while silver *Zan* was worn by middle and lower class women.

The head of *Zan* was rich in design, which presented different shapes such as, flowers, birds, fish, insects and beasts. The exquisite phoenix pattern was commonly used.

• 白玉龙首簪（明）

此簪长11.6厘米，簪身用减地雕法雕一龙，上有篆书若干字，雕工、文字均工雅。

White Jade Hairpin (Ming Dynasty)

This hairpin was 11.6 centimeters in length. The simplified dragon pattern and several seal characters were engraved on this hairpin neatly and orderly.

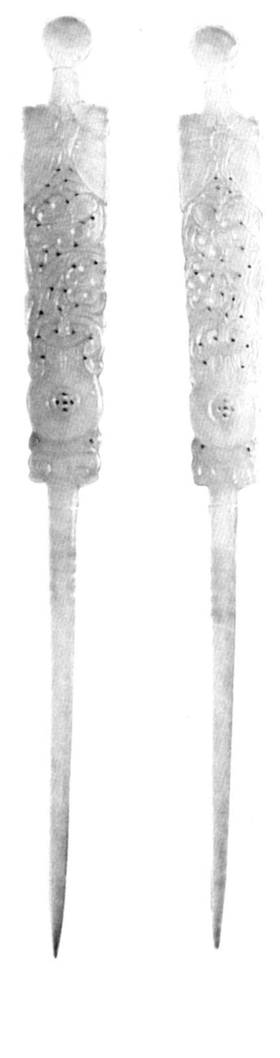

• **花蝶纹玉簪（明）**

在各类发簪中，玉簪一直备受古代女子宠爱，一件上等的玉簪在古代价值连城。

Jade Hairpin with Butterfly Pattern (Ming Dynasty)

Among all kinds of hairpins, the jade hairpin was deeply loved by ancient women. The top grade jade hairpin was priceless in ancient times.

• **金簪（明）**

金簪的变化主要表现在簪首，常见的簪首形状有球形、花卉形，其他还有凤形、鱼形、蝴蝶形、如意形等。

Gold Hairpin (Ming Dynasty)

Gold hairpins varied in the head of the hairpin. There were phoenix-shaped, fish-shaped, butterfly-shaped and *Ruyi*-shaped. The most common ones were spherical shape and floral shape.

镶珠宝灵寿纹金簪（清）

此簪为金质，其上有点翠，錾刻加累丝五个灵芝形花朵，上嵌碧玺等宝石。篆书“寿”字中间嵌东珠一粒，并有点翠的松枝及竹叶点缀其中。簪另一端为镂空累丝长针状。

Gold Hairpin with Inlaid Gems and Longevity Pattern (Qing Dynasty)

It was made of gold inlaid with tourmaline and other gems. Five chiseled ganodermas were made of gold wires. A bead was inlaid on the center of the seal character of "longevity" dotted with pine needles and bamboo leaves. The other part of the hairpin was made of gold wires, hollow and needle-shaped.

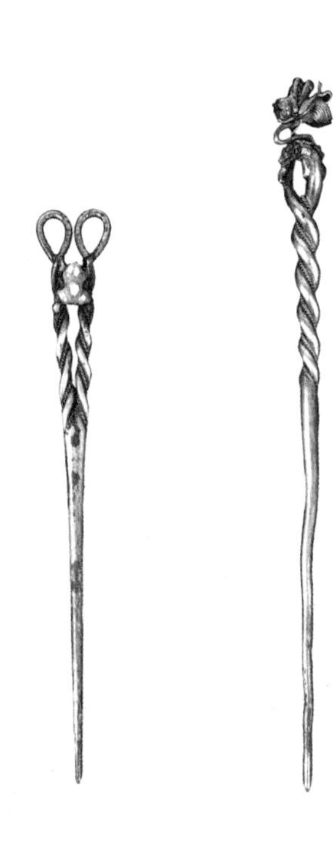

麻花纹银簪（清）

在清代，麻花纹的头簪十分流行，普遍使用于中、下等家庭妇女群体中，制作简单，纹饰规整。

Silver Hairpin with Fried Dough Twist Design (Qing Dynasty)

It was very popular in the Qing Dynasty. Because it was easy to make and the design was neat and orderly, it was widely used among housewives in middle and lower class.

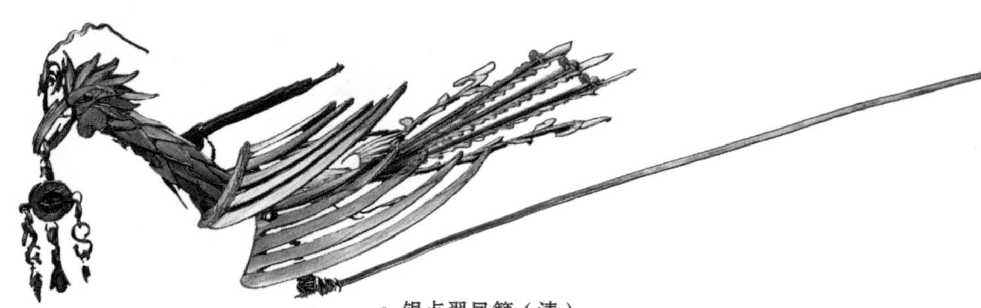

• 银点翠凤簪（清）

Silver Hairpin with Emerald Green Phoenix Pattern (Qing Dynasty)

• 四艺银簪（清）

四艺即琴、棋、书、画，是中国历代文人雅士必备的技艺，是天下太平、生活安逸的象征。此簪中间还有一花瓶，插着牡丹花，寓意富贵平安。

Silver Hairpin with Four Skills Pattern (Qing Dynasty)

Four necessary skills referred to playing musical instrument, playing chess, calligraphy and painting mastered by ancient literatus. It reflected the peaceful world and comfortable life. In the middle of the pattern was a vase holding a peony which symbolized good fortune and safety.

如意纹镏金银簪（清）

清中期以后，如意是官员朝觐用的标识，也是皇族贵戚的馈赠品。此三件如意纹簪采用錾花和镂空工艺，十分精致。

Gilt Silver Hairpins with *Ruyi* Pattern (Qing Dynasty)

These silver gilded hairpins were made by the techniques of engraving and hollowing-out carving. After the mid-Qing Dynasty, *Ruyi* (S shaped ornamental object, a symbol of good luck) was used as present among royal families. Officials used *Ruyi* as the identification to be presented at the royal court.

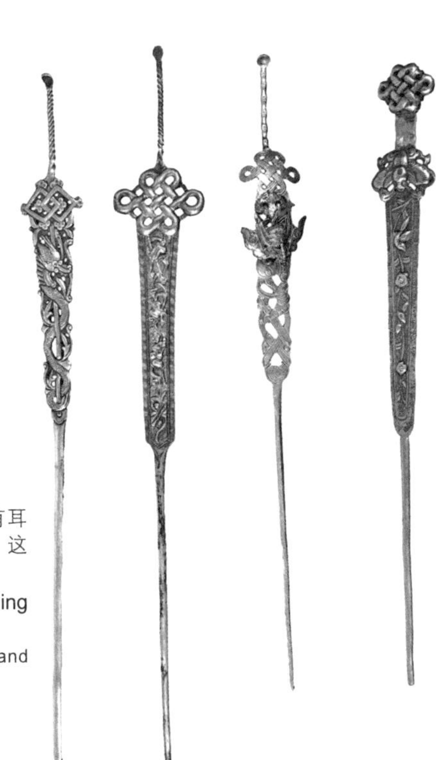

盘长纹耳挖银簪（清）

耳挖簪是一种特殊形制的簪，即簪首前端有耳挖，或实用，或用来做装饰，或二者兼备。这三件耳挖簪都装饰有盘长纹。

Silver Hairpin with Earpick on One End (Qing Dynasty)

These three hairpins were both ornamental and practical so they were a special type of hairpin.

• **金嵌珠连环花簪（清）**

此簪身錾刻连环状，顶端一环弯曲，上嵌一粒珍珠，另一端呈长尖形。

Gold Hairpin Inlaid with Bead

Chains of rings were chiseled on the hairpin with a bead on the top. The first ring was bent and the other end was tapered in shape.

• **翠羽簪（清）**

翠羽簪又称“点翠簪”，是在金银簪架上粘贴翠鸟羽毛而制成。一般为翠绿色，色彩艳丽，配上金边，或镶嵌珠宝玉石，显得富丽堂皇。

Hairpin with Kingfisher Feathers (Qing Dynasty)

Kingfisher feathers with emerald bright color were pasted on the gold and silver hairpin. It was inlaid with gems or gold edge, which looked magnificent.

簪花

在古代，女子还喜欢戴一种特殊的发簪，称为“簪花”。簪花的风习由来已久，魏晋南北朝时，就有簪花的习俗。盛唐时，女子们喜欢梳高髻，还在高耸的发髻上插上花卉。每逢春回大地百花争艳时，唐都长安的仕女们便纷纷加入斗花竞赛，以插戴的花卉最多、最奇异多姿者为花中之魁。由于鲜花容易枯萎，于是人们又想出了用罗绢或彩纸做成假花来代替的方法。假花精巧逼真，经久耐用，深受女子喜爱。

Floral Hairpin

Women in ancient times liked to wear flowers on the head, known as "floral hairpin". The prevalence of wearing the floral hairpin could be traced back to the Wei, Jin and the Southern and Northern dynasties (220-589). In the Tang Dynasty, women like to wear their hair in high buns with flowers. When spring came, ladies in Chang'an, the former capital of the Tang Dynasty, would compete with each other. The one who wore the most beautiful and diverse flowers would rank the first. People started to use silk and colored paper to make artificial flowers, considering that the flower withered easily. Because they were delicate, lifelike and durable, artificial flowers were welcomed by women.

- **《簪花仕女图》周昉（唐）**

《簪花仕女图》中描绘了春日阳光下，几位衣着华贵、雍容闲雅的贵妇，正在赏花游园。她们头上高髻簪花，有的头簪一朵硕大的牡丹花，有的头簪海棠花，还有的插戴素淡的芍药、粉荷。这幅画反映了唐代女子簪花的盛况。

Court Ladies Wearing Flowered Hairpins by Zhou Fang (Tang Dynasty)

It depicts several graceful ladies wandering in the garden to appreciate the beautiful flowers in the spring sunshine. They are dressed luxuriously and wear their hair in high buns. Some pin a large peony on the head. Others wear Chinese flowering apples, plain-colored Chinese herbaceous peony and pink lotus. This painting reflects the prevalence of floral hairpins in the Tang Dynasty.

钗

古代女子的首饰中还有一种发饰，称为“钗”。钗和簪都用来插发，只是簪作成一股，而钗则作成双股，形状像树枝枝丫，也有少数钗作成多股。

早在东汉时期，女子便流行戴钗，常于发间斜插六钗。当时贵族女子的发式以高大为美，人们会在真发中掺入假发，然后用发钗来固定。隋唐时期，女子的发钗不仅品种繁多，而且插戴数目也极多，贵妇们常常着华服、梳高髻，再配以满头的金、银、花钗。明代以后，高髻之风日渐式微，开始流行垂式低髻，钗的形制大多精致小巧。

Chai

Chai was similar to *Zan* and both of them were used to pin hair. *Zan* was single-stranded while the shape of *Chai* was double-stranded, similar to a branch. A few of them were multi-strand.

As early as the Eastern Han Dynasty, *Chai* was popular among women and six-strand *Chai* was commonly worn at that time. The fashionable hairstyle tended to be worn high mixed with wigs and pinned by *Chai*. The Sui and Tang dynasties witnessed a variety of hairpins. Ladies wore decent clothes wearing their hair high in buns matching a number of silver and gold hairpins. The prevalence of wearing hair high in buns gradually declined after the Ming Dynasty. The

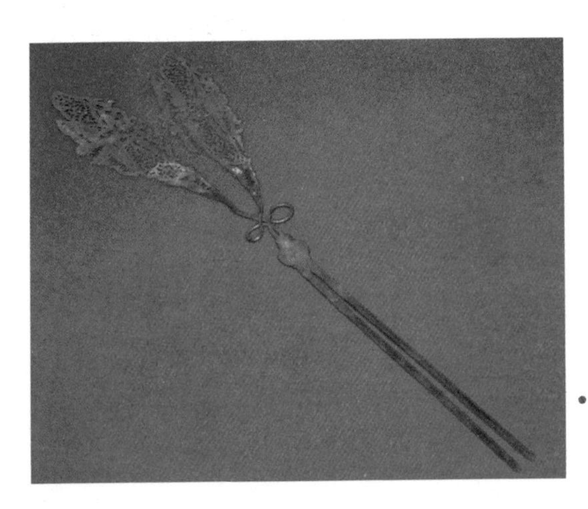

• **蝴蝶纹鎏金银钗（唐）**
Gilded Silver Hairpin with Butterfly Patterns (Tang Dynasty)

钗的安插有多种方法，有的横插，有的竖插，有的斜插，还有自下而上倒插的。所插数量也不尽一致，视发髻需要可插两支或数支，最多的在两鬓各插六支，合为十二支。钗的造型丰富多样，名称也十分形象，多以钗首的造型命名，如燕钗、花钗、凤钗等。制钗的原料

hairstyle of wearing hair in low buns became popular. *Chai* was small but exquisite.

There were a number of ways to wear *Chai*. They could be pinned in a line, in vertical way, in oblique way and bottom-up way. The number of *Chai* to be worn varied based on different hairstyles. Twelve *Chai* could be worn

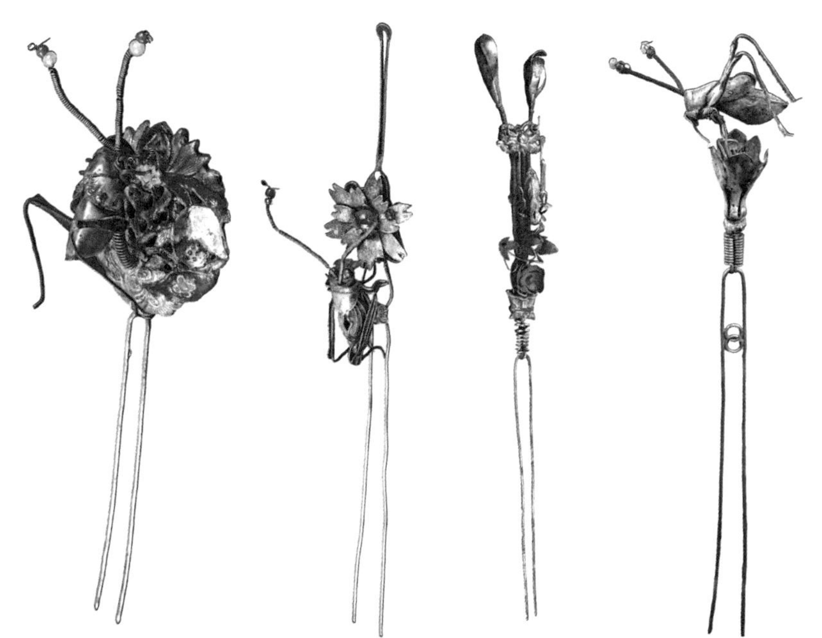

蝈蝈白菜银钗（清）

“蝈蝈”与“哥哥”谐音，俗称“喜叫哥哥”，意在祈求多生男孩。民间常以白菜配蝈蝈，因“蝈”与“官”谐音，“白菜”南方人叫“生菜”，与“生财”谐音，寓意升官发财。

Silver Hairpin with Katydid and Chinese Cabbage Pattern (Qing Dynasty)

In Chinese, the word "katydid" (*Guoguo*) and the word "brother" (*Gege*) are the harmonic tone. The katydid pattern was used to pray for having more sons. According to the pronunciation in southern China, "Chinese cabbage" (Shengcai) and "make fortune" are the harmonic tone. The pattern of matching the katydid with the Chinese cabbage was often used among common people, symbolizing getting promotion and making fortune.

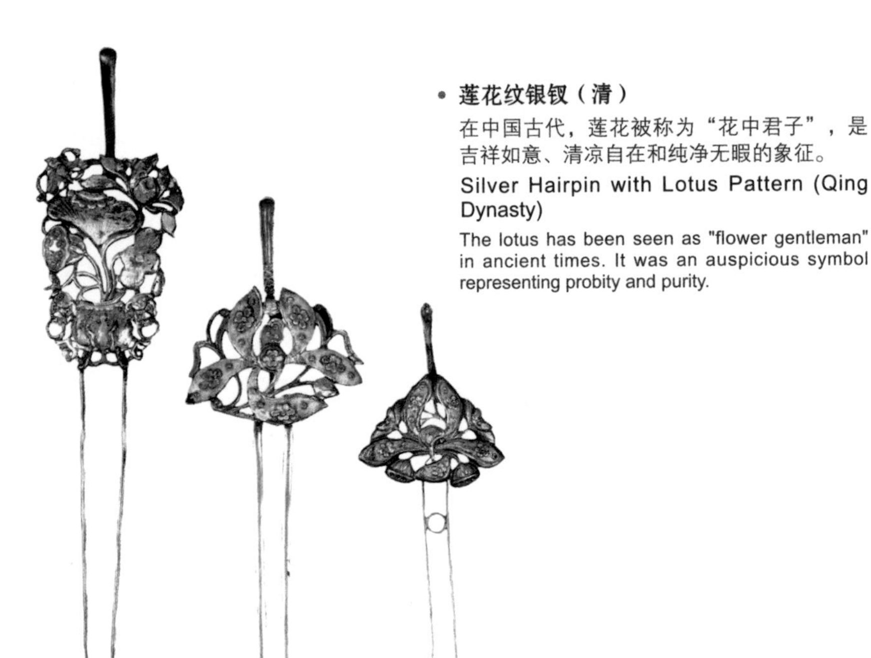

• **莲花纹银钗（清）**
在中国古代，莲花被称为“花中君子”，是吉祥如意、清凉自在和纯净无暇的象征。
Silver Hairpin with Lotus Pattern (Qing Dynasty)
The lotus has been seen as "flower gentleman" in ancient times. It was an auspicious symbol representing probity and purity.

也很多，如金、银、铜、玉、骨、象牙、珊瑚、琉璃等。古代家境贫寒的女子，无钱购买名贵材质的发钗，便只能以荆枝插发，称为“荆钗”。

钗分两股，看似分离，实则有钗头相连，两个钗脚相互依赖，缺一不可。因此古人还将钗看成是一种寄情的物件，恋人或夫妻之间流行一种离别时赠钗的习俗：女子将两股的钗一分为二，一半赠给对方，一半自留，待到他日重见时再合在一起。

on hair at most with six at each side. *Chai* was named after its design such as swallow-shaped *Chai*, floral *Chai* and phoenix-shaped *Chai*. Such precious materials as gold, silver, copper, jade, bone, ivory, coral and colored glaze were worn by the wealthy. What the poor wore was wattles, known as "wattle hairpin".

The design of *Chai* was mostly in two strands with the connection on the top. Because of two-stranded structure, ancient people took *Chai* as a token of love. There was an old custom. Women divided two-stranded *Chai* into two parts keeping a half and giving another half to her beloved. On the day of reunion, two parts were being together.

- **鹤纹银钗（清）**

  中国古人认为鹤是长寿仙禽，具有仙风道骨，常以鹤寿、鹤龄、鹤算等作为祝寿之词。

  Silver Hairpin with Crane Pattern (Qing Dynasty)

  Ancient people believed that the crane was an immortal bird, thus the crane as a symbol of longevity was often used to offer birthday congratulations.

- **蝉纹银点蓝钗（清）**

  蝉纹在古代作为吉祥纹样，有一脉相承、连续不断之意。

  Blue Enameling Silver Hairpin with Cicada Pattern

  As an auspicious pattern, the cicada pattern in ancient China represented never ending.

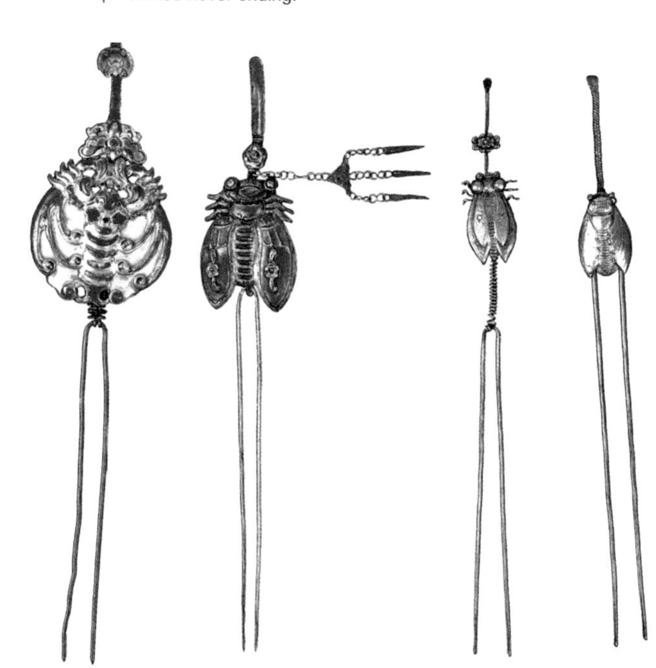

• **菊花纹银点蓝钗（清）**

菊花在中国被誉为“长寿之花”，同时“菊”与“居”谐音，也寓意安居乐业。

Blue Enameling Silver Hairpin with Chrysanthemum Pattern (Qing Dynasty)

The chrysanthemum has been seen as "immortal flower". The word "chrysanthemum" and the word "inhabitation" are the harmonic tone in Chinese so this pattern has the meaning of enjoying a good and prosperous life.

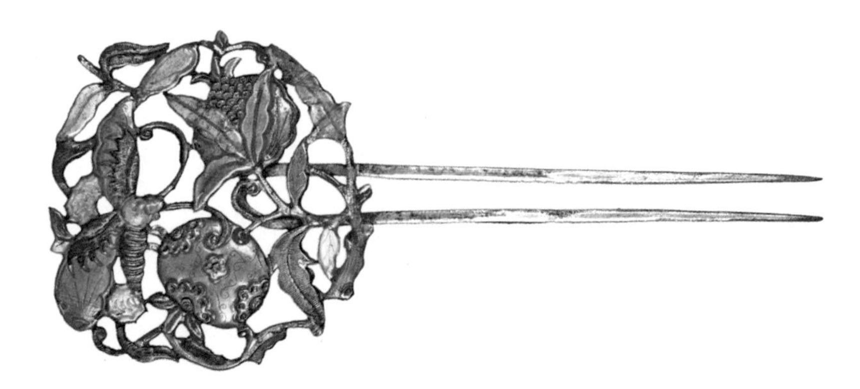

• **石榴纹银点蓝钗（清）**

石榴在中国古代是多子的象征，此钗的石榴纹中又有一只蝴蝶，寓意多子又多寿。

Blue Enameling Silver Hairpin with Pomegranate Pattern (Qing Dynasty)

In ancient Chinese's mind, the pomegranate was a symbol of more children. There was a butterfly on the pomegranate with an additional meaning of longevity.

• **蝶恋花银钗（清）**

以蝴蝶、鲜花组合构成的图案，称为“蝶恋花”。在中国古代，蝴蝶被人视为美好吉祥的象征，“蝶恋花”寓意甜蜜的爱情和美满的婚姻。

Silvery Hairpin with Butterfly and Flower Pattern (Qing Dynasty)

The pattern which was combined the butterfly with the flower was regarded as sweet love and happy marriage. It had an auspicious meaning in ancient people's mind.

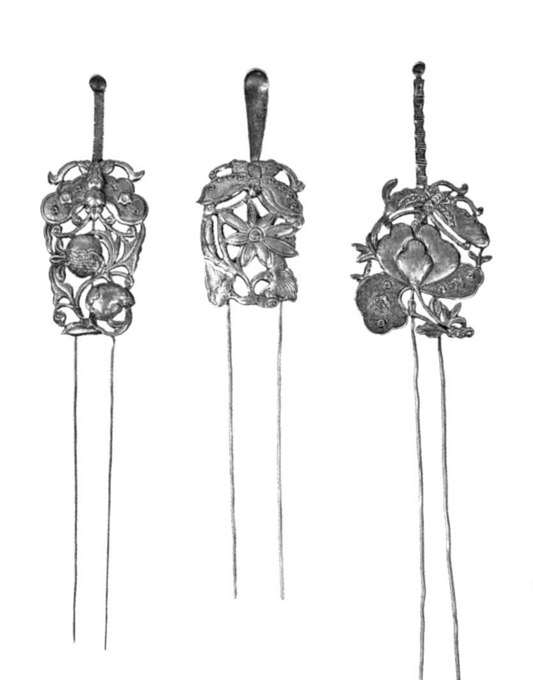

• **金蟾纹四股银点蓝钗（清）**

金蟾在中国民间是一种吉祥物，寓意吉祥富贵。

Four-stranded Blue Enameling Silver Hairpin with Golden Toad Pattern (Qing Dynasty)

Golden toad in Chinese folk-custom is an auspicious animal symbolizing wealth.

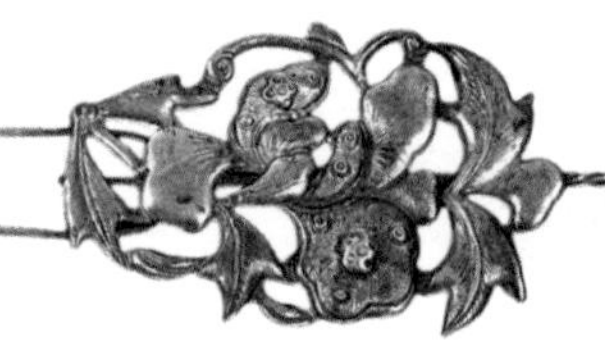

• **牡丹纹银耳挖钗（清）**

牡丹有“花王”之称，并有“国色天香”之誉，故成为富贵和荣誉的象征。牡丹图案被普遍运用在各种吉祥饰品中。

Silver Hairpin with Peony Pattern and Ear pick

The peony has been known as "the king of the flower" enjoying the reputation of "national beauty and heavenly fragrance" hence the peony pattern was widely used.

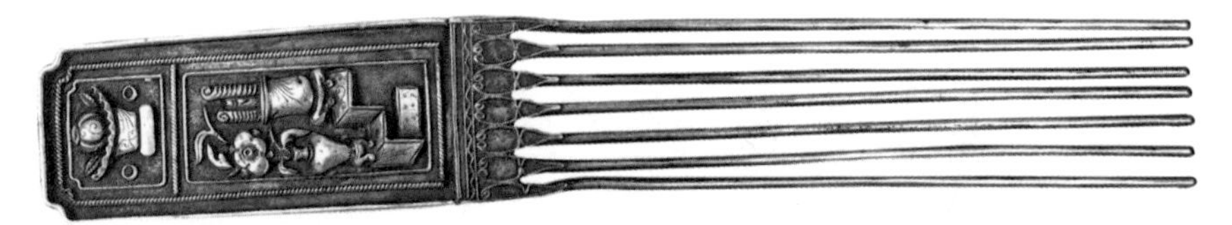

• **博古图七股银钗（清）**

博古图是将国画或器物作为装饰图形的工艺品。该钗装饰很有特色，在一定的范围内留出两个独立的空间，将要表现的器物放进去，形成一个特殊的画面。

Seven-stranded Silver Hairpin with Ancient Pattern (Qing Dynasty)

The pattern showed the ancient artifacts such as Chinese painting, antique and curio. This hairpin was designed to leave room for holding special items. With this ornamental style, the special scene would be presented.

裙钗

裙和钗都是中国古代女子的衣饰，因此常常以“裙钗”或“金钗”来借指女子。如中国古典文学名著《红楼梦》中的“金陵十二钗”，描写了南京（古称“金陵”）十二个最优秀的女子的故事。

Petticoats and Hairpins

Both petticoats and hairpins were women's costume and accessory, hence they were used as the pronoun of women in ancient China. The well-known Chinese classical literature *A Dream of Red Mansions* was also referred to as *Twelve Gold Hairpins*. Twelve hairpins referred to twelve outstanding ladies.

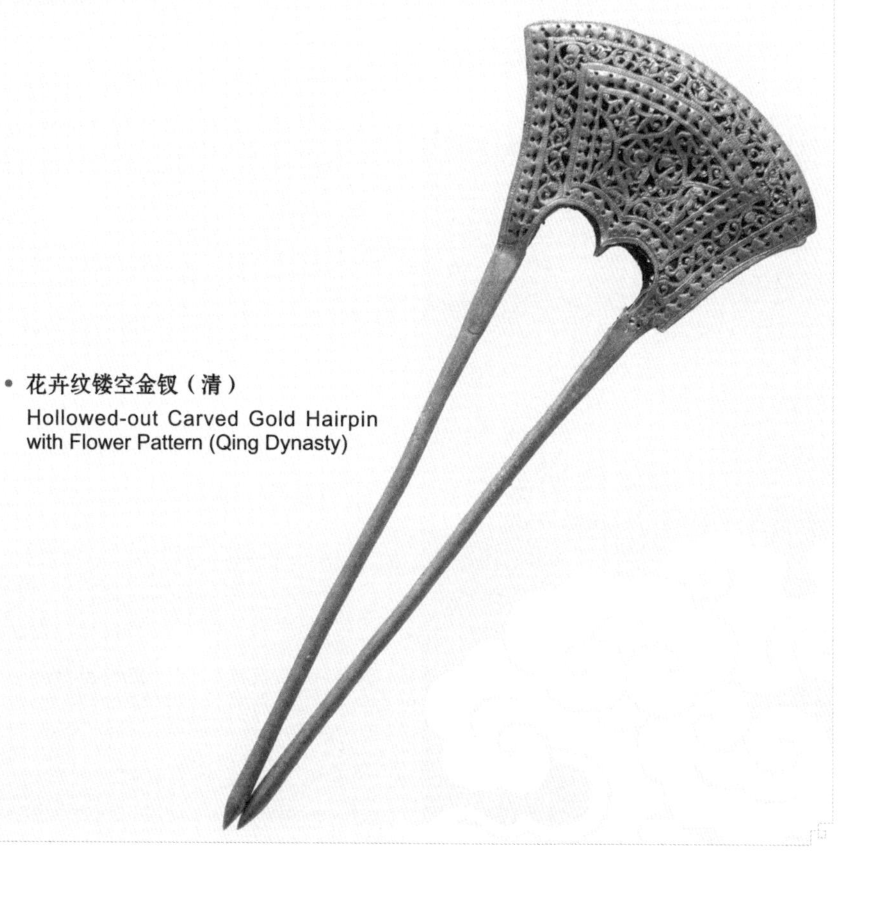

- **花卉纹镂空金钗（清）**
  Hollowed-out Carved Gold Hairpin with Flower Pattern (Qing Dynasty)

步摇

步摇是缀有活动的坠饰的簪或钗，由簪、钗发展而来，其坠饰会随着女子步履的走动不停地摇曳，故得名。步摇大多以金、银、玉等材质制成，其形制与质地都是等级与身份的象征。步摇又被人称为“禁步”，意在约束女子的行为。戴步摇的女子行动要从容不迫，使步摇发出有节奏的声响。

步摇始见于汉代，最初只流行

Buyao (Shaking-while-walking)

The design of *Buyao* evolved from *Zan* and *Chai*. Because there were pendants hanging on it, it swayed with women's each step, hence its name. *Buyao* was made of gold, silver and jade. Its material and shape varied with the social status. *Buyao* was known as "forbidden walking" for the purpose of regulating women's manners. Women had to walk properly so that *Buyao* made rhythmic sound.

Buyao was popular in the court and among aristocrats when it appeared in the Han Dynasty. *Buyao* was so popular in the Wei, Jin and the Southern and Northern dynasties that it was loved by

- **状元游街银步摇（清）**

此步摇的雕饰极为细致，展现了状元高中后骑于马上，前后有人报喜、庆贺的场景，寓意仕途顺利，功成名就。

Silver *Buyao* with the Pattern of Number One Scholar Parading on the Street (Qing Dynasty)

The pattern depicted the scene when he came first in the highest imperial examination riding on horseback, someone announced the good news and congratulated him. It signified that official career went on smoothly and achieved success. This hairpin was engraved exquisitely.

于宫廷与贵族之中。魏晋南北朝时期，步摇最为盛行，花式繁多，民间女子也开始佩戴。到了唐代以后，步摇多为金玉制成、凤鸟口衔的串珠，称为“凤头钗”。除了金质的步摇以外，这一时期还出现了玉、珊瑚、琉璃、琥珀、松石、水晶等珍贵材料制作的步摇。明清时期，步摇再次流行起来，称为“步摇簪”或“步摇钗”。

common women. Therefore, the patterns of *Buyao* became diverse. After the Tang Dynasty (618-907), *Buyao* was made of gold and jade with the design of a bead in phoenix beak, known as "crested hairpin". In addition to gold, it was made of some precious materials such as jade, coral, colored glaze, amber, turquoise and crystal. *Buyao* made a comeback in the Ming Dynasty (1368-1644), known as Shaking-while-walking Hairpins.

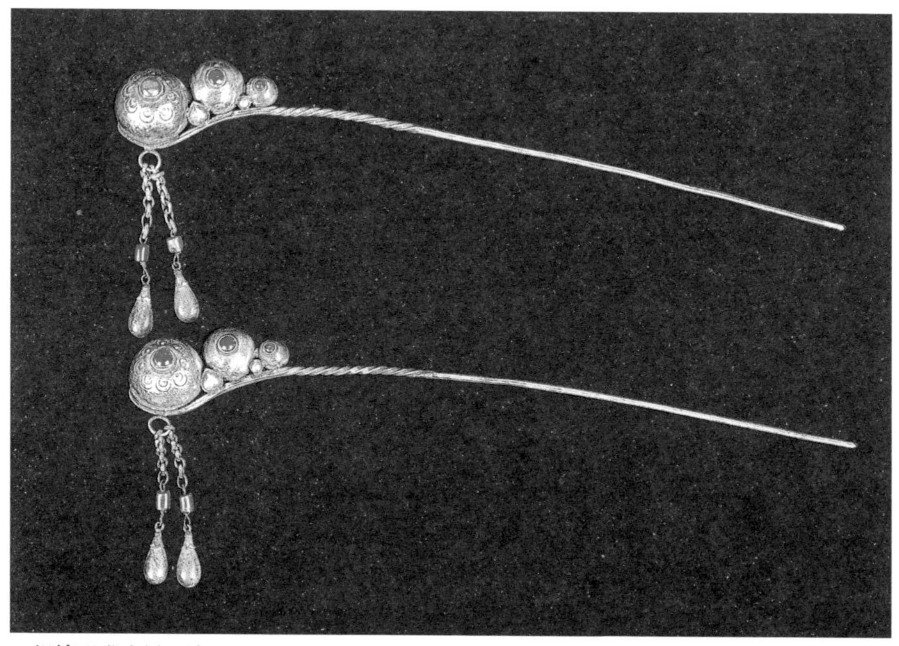

银镶珊瑚步摇（清）

此步摇镶嵌红珊瑚，点缀得体，工艺严谨，十分精美。

Silver *Buyao* Inlaid with Corals

This exquisite *Buyao* was inlaid with corals. The craftsmanship was rigorous and sophisticated.

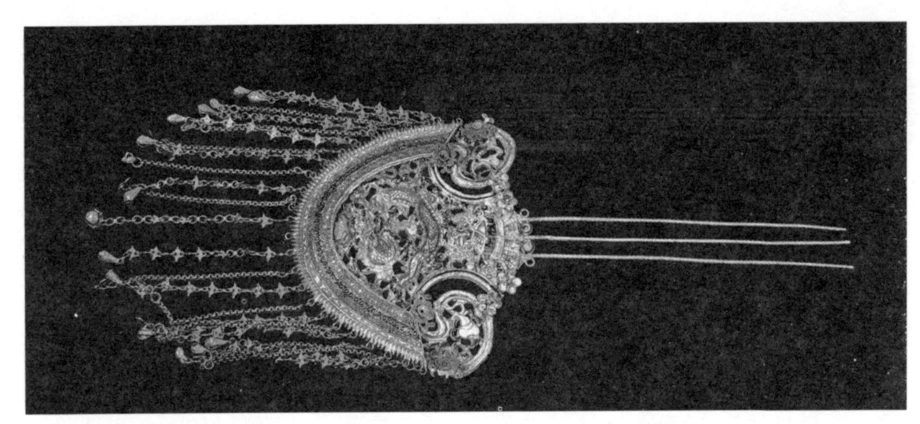

- **珐琅彩扇形银步摇（清）**

珐琅彩的扇形步摇最受清代女子欢迎。“扇”与“善”同音，因此扇形寓意以善为本、以善立德的精神，展开的扇面寓意光芒四射、持续繁荣。

Enamel-color Silver *Buyao* in Fan Shape (Qing Dynasty)

The fan-shaped *Buyao* was welcomed by women in the Qing Dynasty. "Fan" (*Shan*) and "kindness" (*Shan*) are homonyms, so "fan" was used to symbolize kindness. The unfolded fan represented sunshine and prosperity.

- **珐琅彩戏曲人物银步摇（清）**

此步摇工艺十分精湛，錾刻线条圆润流畅，工匠运用高超的技术将人物、亭台楼阁边饰生动传神地呈现出来，就连骑马拍马屁股的动作也能看得很清楚。

Enamel-color Silver *Buyao* with Opera Characters Pattern (Qing Dynasty)

Engraving craftsmanship was so sophisticated that such details as the edging on pavilions, terraces and towers were engraved lively and vividly. Even the motion of slapping the crupper could be seen clearly.

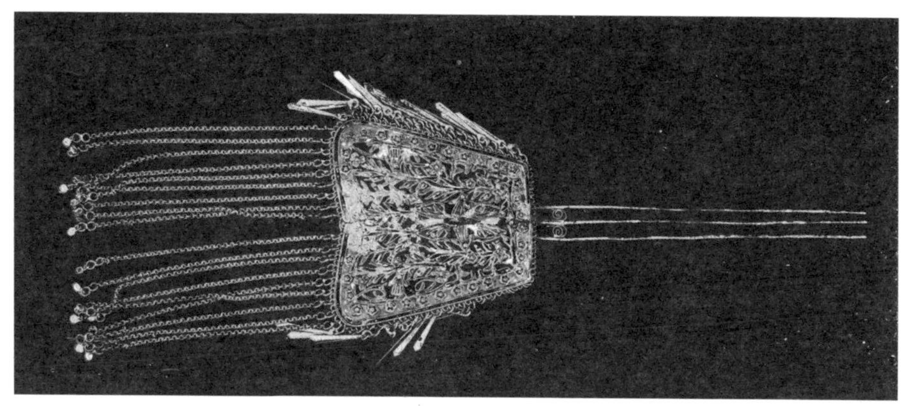

- **珐琅彩扇形花卉纹银步摇（清）**

此步摇构图疏密有致，加上珐琅彩的衬托，使其极富装饰性。钗的前部和两侧各吊有大小不一的银坠，象征连绵不绝。

Enamel-color Fan-shaped Silver *Buyao* with Plower Pattern (Qing Dynasty)

The composition of this pattern was well-arranged. It looked highly decorative against the background of enamel. Different sizes of pendants were hung on the front and both sides of the hairpin symbolizing continuous succession.

- **鱼跃龙门银镀金步摇（清）**

古人用鱼跃龙门比喻高升和行运。此步摇下坠各种吉祥小物件，做工十分精致。

Silver *Buyao* with Carp Jumping Dragon Gate Pattern (Qing Dynasty)

The pattern of the carp jumping dragon gate was a metaphor for getting promotion and having a good luck. A variety of delicate auspicious pendants was hung on this *Buyao*.

中国古代女子的发式

中国古人称头发为“乌云”，称发髻为“蟠龙”。云无定形，变化无穷，瞬息万变；龙有鳞爪，腾云驾雾，姿态万千。“龙”“云”二字，道尽了中国发型的丰富多彩、变化多端，仅唐代段成式的《髻鬟品》就记载了百余种发式。

远古时期，女子的发式简便自然，披发或编发辫，发式并不多。商周时期，少女发辫垂肩，成年女性则把头发盘梳成顶心髻，再横穿一支笄，可以佩戴首饰，也可以戴头巾。自发髻出现，发辫逐渐式微，此后的历朝历代女子皆以梳髻为主。春秋战国时期，成年女子一般梳一条长长的马尾辫，有的在长辫中结成双环，有的以绳系发根。也有的把头发盘成髻，垂直于脑后，发髻后倾，形似银锭或马鞍。秦汉

《女史箴图》中正在梳妆的女子

《女史箴图》是东晋时期的著名绘画作品，作者为顾恺之。此画以写实手法反映了古代贵族女子的形象。画面中一贵妇在铜镜前席地而坐，周围摆放着各种梳妆用具，一侍女正为其理发梳妆。侍女头梳高髻，上插步摇首饰，髻后垂有一髾。这种发式早在汉代就已经出现，魏晋以后再度流行。

The Admonitions of the Court Instructress

It was painted by Gu Kaizhi, a top painter of the Eastern Jin Dynasty (317-420). It depicted the portrait of the ancient aristocratic lady in a realist style. She was sitting on the floor in front of the bronze mirror with a variety of makeup items around her when a maid combed her hair and wore makeup for her. The maid wore her hair in a high bun pinned by hairpins leaving a tuft of hair dangling from the bun. This hairstyle appeared as early as the Han Dynasty. It revived after the Wei and Jin dynasties.

• 《四美图》（宋）

图中描绘了四个盛装的仕女，皆身着襦裙，头戴花冠。仕女的形象、衣饰都仿唐制，但发髻花冠式样又有宋代特色。

Four Beauties (Song Dynasty)

It depicted four beauties in Sunday best. They wore *Ruqun* (a kind of opposite front pieces garment without sash matching with the skirt) and floral ornament. The costumes they worn retained the style of the Tang Dynasty but the floral ornament had the characteristics of the Song Dynasty.

女子以绾髻为主，发式丰富多姿，有文献记载的发髻名称多达数十种。魏晋南北朝时期的发式受外来形式与风格的影响，种类更加繁多。唐代，开放的社会造就了包容的胸怀，唐人大胆创新，在妆饰上追求华丽风雅，发式也随之变得新颖丰富。宋代女子承继晚唐遗风，以高髻为时尚。同时宋人喜大髻，创造了盘龙髻，即将头发梳绾盘曲至头顶，多用假发衬入髻内，再插以簪钗，饰以珠翠，远远望去，犹如一条待时而飞的潜龙。明初妇女的发髻基本上保持宋元时期的式样。明中期以后，发髻多梳于头顶，有的偏向一侧，发髻有一个或多个不等，造型各异，变化多端，且以松散蓬松为时尚。清代女子的发式分为满、汉两种。清初，满、汉两种发式还各不相同，后来开始相互影响，都发生了明显变化。满族妇女的典型发式为两把头、如意头、架子头、旗髻等。汉族妇女的发髻在清初时基本上沿用明代发式。

- **唐代仕女陶俑**

  唐代女子的发式非常丰富，这些陶俑即展现了当时几种常见的发髻。

  Pottery Figurines of Maids in the Tang Dynasty

  Women's hairstyles of the Tang Dynasty were diverse. These pottery figurines showed several common hairstyles in the Tang Dynasty.

Hairstyle of Ancient Women

Hair was referred to as "dark clouds" and hair buns were known as "curled-up dragon". Because of the endless variations and various shapes, the close analogy with the diverse hairstyles was made. Several hundred hairstyles were recorded in the book of *Ji Huan Hairstyle* written by Duan Chengshi in the Tang Dynasty.

In ancient times, women's hairstyle tended to be natural and simple with hair falling over the shoulder or wearing braids. In the Shang and Zhou dynasties, girls wore braids down the

- **《千秋绝艳图》中的明代女子**

  明代女子的发髻造型各异，有的形似牡丹、荷花，有的状如盘龙、鹅胆，有的像钵盂。发髻的名称有桃心髻、鹅胆心髻、堕马髻、牡丹髻、盘龙髻等。

  Picture of Beauties Depicted the Beauties in the Ming Dynasty

  Women's hairstyles of the Ming Dynasty were unique. The hair buns were lotus-shaped, peony-shaped, crouching-dragon-shaped and alms bowl-shaped, hence their names.

shoulder while women wore the bun pinned by a hairpin decorated with jewels and scarf. Ever since hair buns appeared, braids had gradually fallen into disfavor. Women mostly wore buns in the following dynasties. This tendency had changed since the Spring and Autumn Period and the Warring States Period. Women started to wear their hair in ponytails. Some tied their ponytails in loop shape. Others tied their hair with a ribbon along with the hairstyle of wearing buns leaning back similar to the shape of saddle or silver ingot. The document recorded that there were dozens of different hair buns in the Qin and Han dynasties. Affected by foreign culture, the number of hair styles increased greatly. Thanks to the open vibe in the Tang Dynasty, hairstyles tended to be more creative and accessories became more fashionable. Women in the Song Dynasty preferred wearing their hair in high buns. This hairstyle was passed down from the late Tang Dynasty. Crouching Dragon Bun was created in the Song Dynasty. It was worn on the top of the head added to wig, pinned by the hairpin and decorated with beads. Seen from the distance, it looked like a crouching dragon ready to fly, hence its name. Women's hairstyle in the early Ming Dynasty remained as it was in the Song Dynasty. After the mid-Ming Dynasty, hair buns were worn on the top of the head or on one side. The number of buns varied with the style. Loose hairstyle was fashionable at that time. Women's hairstyles of the Qing Dynasty (1616-1911) were fallen into Han and Manchu. Two styles were totally different in the early Qing Dynasty but they changed greatly with interaction with one another. The typical Manchu hair style included double *Batou*, *Ruyi* (an S-shaped ornamental object symbolizing good luck), *Jiazhi* bun and Manchu bun. In the early Qing Dynasty, Han women's hairstyles basically followed the hairstyles in the Ming Dynasty.

扁方

扁方是清代满族女子特有的首饰，一般呈扁平的“一”字形，用来固定满族女子的两把头发式。扁方呈扁平状，通常以金、银、玉、玳瑁、沉香木等材料制成，上面镶嵌珠玉或刻各种花纹，使用时插于发髻之中。

Prolate Hair Ornament (*Bianfang*)

Prolate hair ornament was a unique accessory for Manchu women. It was in the shape of a straight line. Prolate hair ornament was made of gold, silver, jade, tortoiseshell and aloewood inlaid with beads or engraved with various patterns. It was flat in shape to do Manchu women's hairstyle.

• 头插扁方的满族女子（清）
A Manchu Woman with Prolate Hair Ornament (Qing Dynasty)

清代的王室贵族妇女所戴的扁方从材质到样式都堪称精美绝伦。在扁方仅一尺长的窄面上，工匠能雕刻出栩栩如生的花草虫鸟与亭台楼阁等精美图案。贵族女子们戴扁方时，会将两端的花纹露出，极具装饰性；有的还会在扁方上缀挂丝线缨穗以限制脖颈的扭动，使行动有节，增添女人端庄的仪态之美。

扁方也有很小的，但小型扁方一般只在特殊场合时佩戴。如遇丧事，妻子为丈夫戴孝，就需要将头

The design and material of the prolate hair ornament for the royal women were exquisite. Lifelike patterns such as flowers, grass, insects, birds and landscaping were engraved on a narrow plate roughly about 3.33 centimeters in length. When it was worn by the royal women, they liked to expose decorative patterns on both ends. Some would hang silk tassels on the prolate hair ornament. Wearing this kind of hair ornament, women were restricted to turn their heads randomly so as to make them look graceful.

发编成两条辫子，辫梢不系头绳，头顶上插一个三四寸长的白骨小扁方。如果是儿媳为公婆戴孝，则要横插一个银或铜质的小扁方。

The small-sized prolate hair ornament was worn on special occasions. When the wife wore mourning for her husband, she had to wear two braids pinned by a white prolate hair ornament which was made of bone material and about 10 centimeters in length. Wearing mourning for her parents-in-law, she would wear a prolate hair ornament made of silver or bronze.

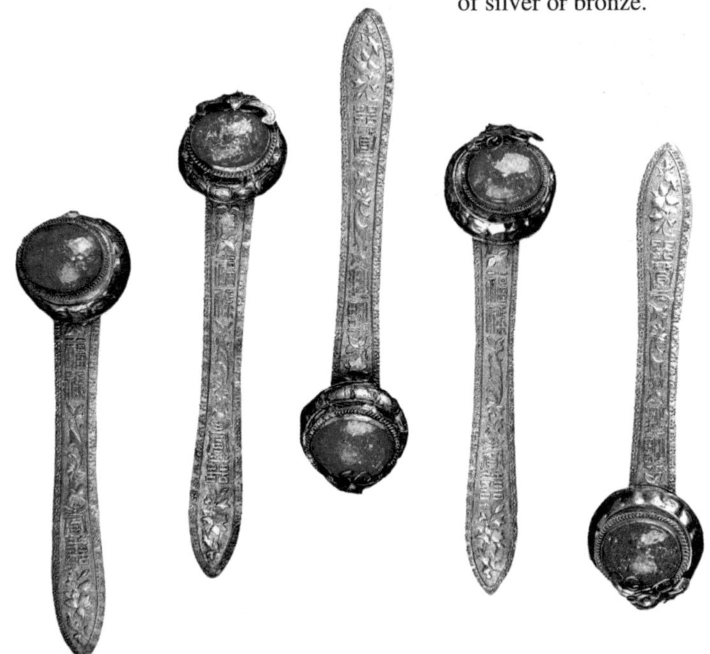

- **银镏金镶碧玺扁方（清）**

  这组扁方镶有碧玺，上有“福禄寿喜”四字，錾花工艺细致，造型花纹边饰均很饱满，再加上镏金工艺，更显华贵之气。

  Gilded Sliver Prolate Hair Ornaments (Qing Dynasty)

  They were *Bixi* (tourmaline) with four Chinese characters meaning happiness, official career and longevity inlaid on it. The edging of the floral pattern was engraved exquisitely. Gilding made them look gorgeous.

- **白玉嵌宝石扁方（清）**

  White Jade Prolate Hair Ornament Inlaid with Gemstones (Qing Dynasty)

- **珐琅彩龟背纹银扁方（清）**

  龟能耐饥渴，寿命极长，因此成了长寿的象征。在中国古代，龟蕴含着天地间神秘莫测的内容，古时候人们用龟甲来占卜以断吉凶。因此，龟背纹是中国传统吉祥纹样之一。

  Enamel-color Prolate Hair Ornament with Tortoiseshell Pattern (Qing Dynasty)

  The turtle is seen as a symbol of longevity because it can endure hunger and thirst. In ancient times, Chinese people used tortoiseshell to divine because the turtle was seen as a mysterious creature. Tortoiseshell pattern is one of the traditional auspicious patterns in China.

- **蝴蝶纹双喜金扁方（清）**

  Prolate Hair Ornament with Butterfly Pattern and Character "Double Happiness" (*Shuangxi*) (Qing Dynasty)

珐琅彩暗八仙纹银扁方（清）

八仙是中国古代民间著名的八位神仙，暗八仙纹即指八仙手持的八件宝物，包括玉笛、芭蕉扇、宝葫芦、荷花等。传说八仙佩的宝物各有神通，民间常以暗八仙为护身符，求平安吉祥。

Enamel-color Silver *Bianfang* with Pattern of Eight Immortals' Eight Treasures (Qing Dynasty)

Eight immortals were well-known in ancient Chinese folklore. Their eight treasures referred to jade flute, palm-leaf fan, the magic gourd, lotus, and so on. They were used as amulets to pray for safety and propitiousness.

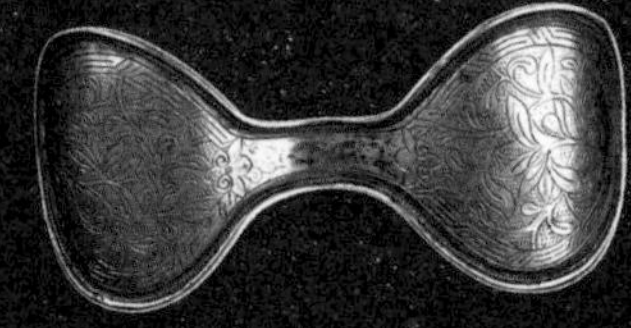

桃形银扁方（清）

桃形扁方是一种样式特殊的扁方，两边为桃形，两头作为装饰的重点，花草人物图案随形而设。

Peach-shaped Silver *Bianfang* (Qing Dynasty)

Peach-shaped prolate hair ornament was a special style. Both ends were peach-shaped decorated with flowers, grass, figures and other patterns.

- 珐琅彩吉祥花卉纹银扁方（清）

这组扁方装饰有牡丹花、梅花、荷花等纹样，从枝到叶构图疏密有致，密而不乱，非常有层次，给人一种吉祥富贵的美感。

Enamel-color Silver *Bianfang* with Auspicious Flower Patterns (Qing Dynasty)

The patterns included peony, plum blossom and lotus. The composition of branches and leaves was well-arranged giving you a sense of beauty.

> 耳饰

耳饰在中国传统饰品中占有重要的地位。早在战国时期，女子就有了穿耳戴环的风俗，当时的耳饰是一种有缺口的环形玉，称为“玦”。女子戴耳饰成为普遍风气始于宋代，此时的耳饰大都用金银玉石打磨制成，多用仿生式造型，如弯月式耳环，上面装饰着蜂蝶花果等纹样，轻巧精美。明代以后，女子穿耳戴环已十分平常。耳环的材料和造型更加丰富，以金、银、玉、珊瑚、玛瑙、珍珠、贝壳、竹木等材料制成各种耳环、耳塞、耳坠，耳饰上的镶嵌之物也极尽华贵而繁复。清代戴耳饰已成风尚，多为耳坠，富贵之家的女子一般都有数十个耳饰，佩戴时看季节及场合而定，还须与衣服的颜色配套，为

> Ear Ornaments

Earrings played an important role in traditional Chinese accessories. The custom of wearing earrings was traced back to the Warring States Period. The prevalence started in the Song Dynasty when most of earrings were made of gold, silver and jade. Earrings were polished to simulate the shape of crescent decorated with beautiful patterns such as bees, butterflies, flowers and fruits. After the Ming Dynasty, wearing earrings became a common practice. Along with gold, silver, and jade, other materials emerged such as coral, agate, pearls, shells and bamboo. They were used to make earrings and eardrops inlaid with gemstones, luxurious and beautiful. Women in the Qing Dynasty most wore eardrops. Women in wealthy family had dozens of eardrops to match what they wore on different occasions. The

• 龙纹玉玦（西周）
Jade *Jue* with Dragon Pattern (Western Zhou Dynasty)

了戴卸方便，一般无须取下耳环，只要更换底下的坠饰即可。

在中国古代有不少一物多名的情况，耳饰也是如此。其称呼因材质、形状和构造的不同而多种多样：形状大的叫“耳环”，形状小的叫“耳塞”；耳环下面缀以垂饰的称“耳坠”；通体以玉或琉璃等为材料，圆筒形、筒腰收缩、一端或两端宽大呈喇叭状的耳环，称作“珥”；耳环下系珍珠的称作“珰”。明清时期还流行一种小巧、淡雅、简洁的耳环，叫作“丁香”。

pendants could be removed from the earrings and replaced with new styles for convenience's sake.

In ancient China, a number of different names were given to the same object. So do ear ornaments. They were named after the material and the shape. Some earrings were made of one piece of jade or one piece of colored glaze named *Er*. They were barrel-shaped in general but narrow in the middle. Other styles were horn-shaped on either end or on both ends. The ear ornament decorated with pearls was named *Dang*. A kind of small-sized and simple ear ornament was popular in the Ming and Qing dynasties named "*Dingxiang*".

• **累丝葫芦形金耳坠（明）**

葫芦耳坠是明代耳环造型中的代表。以金丝弯制成钩状，在金丝的一端，穿上两个玉珠或金珠，小珠在上，大珠在下，两珠上面再覆一片金制的圆盖，整个造型就像一个葫芦。珠子有的做成实心，有的镂空，显得小巧玲珑。

Filigree Gold Gourd-shaped Earrings (Ming Dynasty)

Gourd-shaped earrings as the representative of earrings in the Ming Dynasty were made of the bent gold wires. There were two jade beads or gold beads on one end with small beads on the top of the big beads. Two beads were covered with a round gold cover. Some of the beads were solid and some of them were hollow, small but exquisite.

• **金嵌珠翠耳坠（清）**

Gold Eardrops Inlaid with Jadeites (Qing Dynasty)

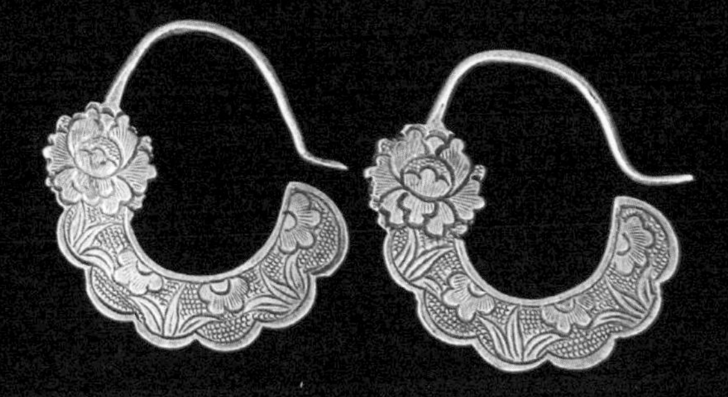

• 兰花纹银耳环（清）
Silver Earrings with Orchid Pattern (Qing Dynasty)

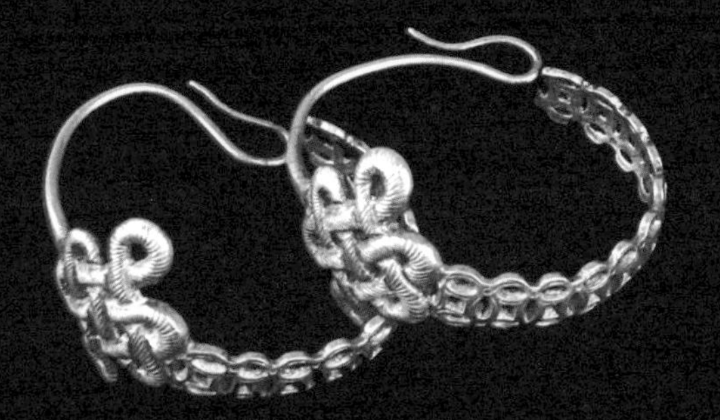

• 盘长纹银镏金耳环（清）
Gilt Silver Earrings with Coiled Pattern (Qing Dynasty)

> 颈饰

古人最早佩戴的颈饰是用大自然赐予的材料——石珠、树子、兽齿、鱼骨、贝壳等穿组而成的，其中以兽齿的数量为最多。很多出土的颈饰都经过磨制、打眼，有的还被矿物染过颜色。

项链是最常见的颈饰，最初的项链是将一串串珠形器、管形器串在一起，两端系住，成为一圈。在

> Neck Ornaments

The neck ornaments worn by ancient people were made of natural materials in cluding stones, animal teeth, fish bones and shells most of which were animal teeth. A number of unearthed neck ornaments were polished, bored and even dyed by minerals.

The original necklace was threaded by a string of beads and tube-shaped ornaments. The necklace was the most common neck ornament. A great number

• 河北邯郸出土的贝壳颈饰（新石器时代）
Unearthed Shell Necklace in Handan, Hebei Province (Neolithic Age)

• 北京房山出土的玛瑙绿松石玉串项链（西周）
Unearthed Necklace Threaded by Agate, Turquoise and Jade in Fangshan District, Beijing (Western Zhou Dynasty)

新石器时代的墓葬中就曾发现过许多骨珠项链、绿松石项链和玉管项链。汉代以后，颈饰的形状已十分丰富，工艺也极精美，多用金、银、玉、玛瑙、水晶、珍珠等材料制成。唐代的颈饰除了沿袭前代的样式和材质，还开始盛行戴璎珞。璎珞本是佛像脖颈间的装饰，随着民间女子佩挂佛珠成为风尚流行开来，逐渐发展成为普通的颈饰。到

of bone bead necklaces, turquoise necklaces and jade tube-shaped necklaces were unearthed from the Neolithic tombs. After the Han Dynasty, the craftsmanship of neck ornaments was exquisite. The shapes and materials became diverse such as silver, gold, jade, agate, crystal and pearls. Neck ornaments of the Tang Dynasty followed the style of previous dynasties. The neck ornament which was used to decorate the statue of Buddha was popular in the Tang Dynasty. It prevailed among the women at firse and then it was widely worn among common people. During the Ming and Qing dynasties, the style of the necklace became more diversified and craftsmanship was superb.

In addition to the necklace, ancient people had the custom of wearing chaplet. Chaplets were forged by gold, silver, bronze and other metals. Some of them were carved by a whole piece of jade. Auspicious longevity lock and *Ruyi* pendant were hung on the chaplet during the Ming and Qing dynasties.

了明清时期，项链的样式更加丰富，制作工艺也更加精湛。

除了项链，古人还有戴项圈的习俗。项圈有用金、银、铜等金属锻制的素圈，也有用整块美玉雕制的玉圈。明清时期还盛行在项圈上缀有长命锁、如意等坠饰。

• 玛瑙水晶项链（西汉）
Agate and Crystal Necklace (Western Han Dynasty)

• **一品清廉银项圈（清）**

“一品”指中国古代最高的官阶，而“青莲”与“清廉”同音。“一品清廉”表达了老百姓对公正廉洁的官员的祝颂与企盼。该银项圈采用錾花工艺制成，装饰纹样重点处理在项圈上，花卉、人物、万字不到头等纹样，工艺十分精良。

Chaplet with *Yipin Qinglian* (Top Rank Officer Clean-handed as Lotus) Pattern (Qing Dynasty)

The first rank official referred to the highest rank in ancient China. In Chinese, "lotus" and "free from corruption" are homonyms. It expressed common people's wish for fair and honest officials. Such patterns as flowers, figures and swastika were engraved on the chaplet. Craftsmanship was sophisticated.

> 手饰

手饰即戴在手上的装饰品，在中国有着悠久的历史，种类较多，主要包括手镯、手链、臂钏、戒指、扳指等。

手镯

手镯在中国古代称为“腕环”，链状的称为“手链”。早在新石器时代，人类就开始戴陶环、骨环、石镯等，其形状有圆管状、圆环状，也有两个半圆形环拼合而

> Hand and Arm Ornaments

A variety of hand and arm ornaments appeared in China's time-honored history, such as bracelets, armlets, rings and thumb rings.

Bracelet

The ancient referred to bracelets as "loop on the wrist". As early as the Neolithic Age, people started to wear pottery loop, bone loop and stone bracelet which included round tube shape, circular shape and two semi-circles. The bracelet at

• 绿松石手链（商）
Turquoise Hand Chain (Shang Dynasty)

成的。这一时期的手镯已具有了一定的装饰性，不仅表面磨制光滑，而且有些表面还刻有简单的花纹。商周时期，手镯多为玉制，此外还出现了金属手镯。汉代以来，手镯被用来赏赐有功之臣，不限男女，均可使用，再后来逐渐成为女子的首饰或小孩的贴身饰物。隋唐以后，女子戴手镯已经很普遍了，手镯的材料和制作工艺也有了极大的发展。到了明清时期，金银镶宝石手镯盛行不衰，成为当时最流行的手镯。

手镯不仅仅起装饰作用，它还是一种吉祥物。古代民间女子出嫁时，手镯是最起码的订婚礼物。在古代的文学作品中，常有女子以手镯相赠恋人的描写。

that time was polished and carved out simple patterns. In the Shang and Zhou dynasties, most of bracelets were made of jade. Gold bracelets appeared at that time. Bracelets were used to award to meritorious officials so both men and women were entitled to wear. Bracelets were gradually worn by women and children. Women wearing bracelets became common practice in the Tang Dynasty (618-907). There was a great development in the material of bracelets and making process. Gold and silver bracelets were popular in the Ming and Qing dynasties (1368-1911).

The bracelet was more than an ornament. It was one of the auspicious presents for engagement. The plot of ancient literature often showed that a girl presented her beloved one with the bracelet.

金手链（春秋）

Gold Hand Chain (Spring and Autumn Period)

• **连珠纹金手镯（清）**

连珠纹是用小圆圈作横式排列而产生的纹样，最早出现在青铜器上。

Gold Bracelet with a Line of Bubbles (Qing Dynasty)

This pattern first appeared on bronze vessels.

• **伽南香木嵌金珠寿字手镯（清）**

Qienan Sandalwood Bracelet Embedded in Gold Beads Forming the Characters of "Longevity" (Qing Dynasty)

• **金镶九龙戏珠手镯（清）**

此镯以金栏划分成九格，每格中各錾一团龙，龙口衔珍珠各一。手镯边沿錾刻海水纹，做工精致，具有很强的浮雕效果。

Gold Bracelet Inlaid with the Pattern of Nine Dragons Playing with Beads (Qing Dynasty)

This bracelet was divided into nine bars with nine dragons. Each dragon held a pearl in its mouth. The edge was engraved with sea wave pattern. Exquisite workmanship showed a strong relief effect.

• **金镂空古钱纹镯（清）**

此镂空古钱纹镯为清代后妃佩戴的首饰，是以细金丝条捶击成一个古钱形并将其连接成镯，首尾接合处几乎天衣无缝。虽光素无纹饰，但由一个个金钱形组合成的镂空钱纹镯，在清代后妃所佩戴的诸多首饰中也属罕见。

Hollow-out Carved Gold Bracelet with Old Coin Pattern (Qing Dynasty)

This bracelet was for Empresses of the Qing Dynasty. The old coin pattern was made by gold wires connected with one another without seeing any seals. It looked plain but the pattern and craftsmanship was rare to see among all accessories for Empress and concubines of the Qing Dynasty.

- **镀金银手镯（清）**

此手镯以模压工艺制成，先铸刻钢模，然后将银料打成条状，放入钢模内锻压，纹饰清晰。这种银手镯在清代十分流行，属姑娘出嫁时必备之物。这只银手镯不但花色多姿还有吉祥文字，纹饰也很规整。

Gold-plated Silver Bracelet (Qing Dynasty)

This bracelet was made by mold. The silver ingot was forged into strips and stamped patterns in the permanet mold. Varied floral patterns and auspicious characters were stamped in order. It was an essential accessory for girls on the day of wedding ceremony because this kind of bracelet was popular in the Qing Dynasty.

- **双股藤镀银手镯（清）**

此手镯用双股藤围成圈，接口处用银包镶。黑藤配白银，色彩对比十分强烈。两者质地粗细变化也非常微妙。

Double-stranded Rattan Bracelet Sealed by Silvering (Qing Dynasty)

This bracelet was made of looped double-stranded rattan. The interface was covered by silvering. The rattan and silver provided a great contrast in color and texture.

- **象牙手镯（清）**

Ivory Bracelets (Qing Dynasty)

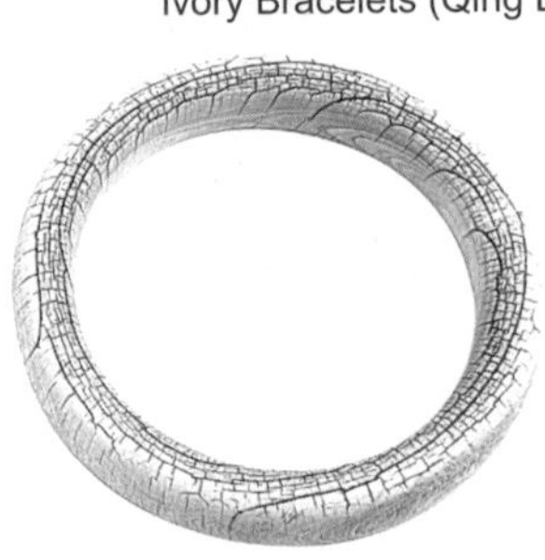

• **单股藤包银手镯（清）**

此手镯用藤围成圈，接口处用银包镶，包银部分充分展示了錾花工艺特色。手镯上錾刻有牡丹纹、福寿纹、卷草纹、八仙纹等纹样，丰富多彩，是人工雕刻与自然纹理的巧妙结合。

Single-stranded Rattan Bracelet Sealed by Silvering (Qing Dynasty)

This bracelet was made of looped double-stranded rattan. The interface was covered by silvering. Silver part made the most of chiseling. Diverse engraved patterns included peony, "happiness" (*Fu*) and "longevity" (*Shou*) characters, curly grass, eight immortals, etc. The design integrated natural textures with artistic carving skillfully.

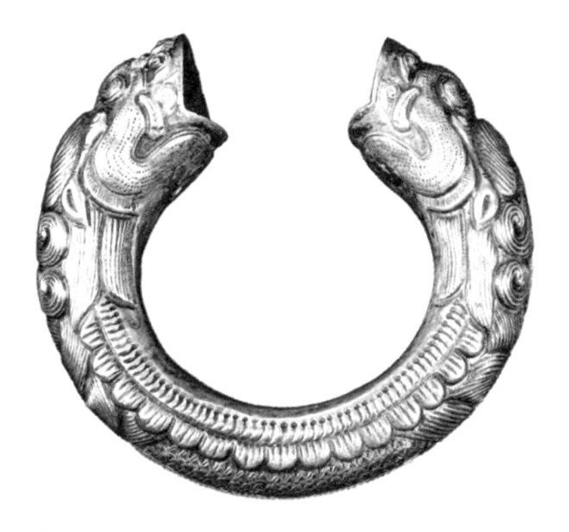

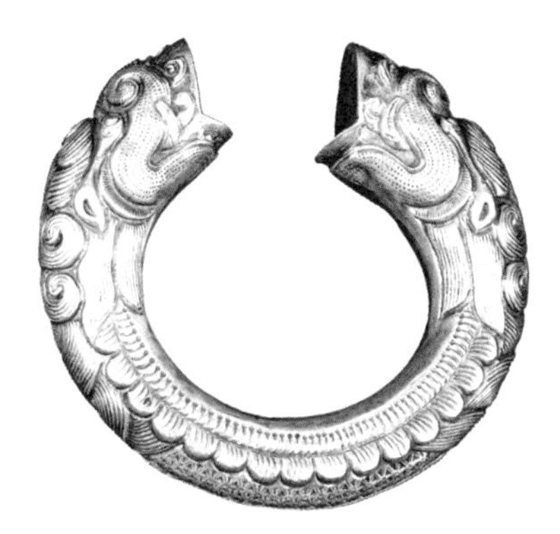

• **龙头纹银手镯（清）**

龙是神圣吉祥喜庆之物，是权威和尊贵的象征，被历代皇室所御用。

Silver Bracelet with Dragon Heads (Qing Dynasty)

The dragon was seen as sacred and auspicious creature symbolizing authority and dignity, thus dragon pattern was for the royal's exclusive use.

- **翡翠手镯（清）**

  翡翠又称“翠玉”，是玉的一种。中国人喜欢佩戴玉镯，因为玉镯名贵且具有装饰作用。

  Jadite Bracelet (Qing Dynasty)

  Chinese people like to wear jade bracelet because it was commonly believed that precious jade bracelet can be used for decoration.

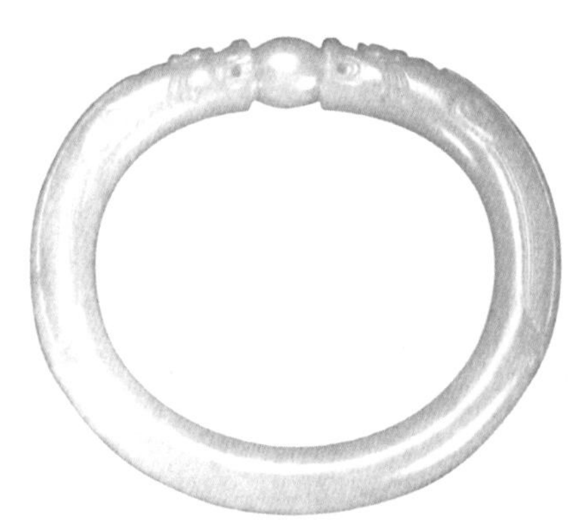

- **二龙戏珠白玉手镯（清）**

  龙凤纹的玉镯主要是宫廷用玉镯，在玉镯上雕龙凤纹作为装饰，取“龙凤呈祥”之意。

  White Jade Bracelet with the Pattern of Two Dragons Playing with a Pearl (Qing Dynasty)

  The bracelet with the dragon and phoenix pattern was exclusively for the royal court symbolizing prosperity.

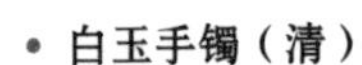

- **白玉手镯（清）**

  此手镯玉质洁净细腻，润如凝脂。表面不加纹饰，打磨得十分光滑。

  White Jade Bracelet (Qing Dynasty)

  The texture of this jade bracelet was clear and neat. It was well-polished without any pattern.

臂钏

臂钏又称为“臂镯”或“臂环”，是一种套在上臂的环形首饰。大多数臂钏都有一个开口处，可以方便地变换尺寸，十分适合手臂粗的女子佩戴。

古代女子戴臂钏的历史十分久远，从西汉开始盛行，样式很多。“臂钏”一词最早出现在隋唐。唐代的女子一般将手帕拴在右腋下的纽扣间，或是把它折叠成为方胜的样子约束在臂钏中。到了清代，由于贵族女子都穿旗袍，所以极少戴臂钏。

Armlet

The armlet was an annular ornament worn on the upper arm. Because most of armlets had an opening which was used to adjust according to the wearer's size, they were suitable for women who had stout arms.

The armlet prevailed in the Western Han Dynasty when there were various styles. The word "armlet" first appeared in the Sui and Tang dynasties. In the Tang Dynasty, women usually tied the handkerchief on the button of the right armpit or folded the handkerchief in the shape of two overlapped diamonds and put it into the armlet. Aristocratic women of the Qing Dynasty wore chi-pao and the armlet fell into disfavor.

• 玉臂钏（商）
Jade Armlet (Shang Dynasty)

• **金臂钏（明）**

这种臂钏称为“跳脱”，如弹簧状，盘拢成圈，少则三圈，多则十几圈，两端用金银丝编成环套，用以调节松紧。这种臂钏可戴于手臂部，也可戴于手腕部。

Gold Armlets (Ming Dynasty)

They could be worn either on the wrist or on the arm varied from three pieces to ten pieces. Chain hooks made by silver or gold wires on both ends were adjustable.

• **银镀金臂钏（清）**

Gold-plated Silver Armlet (Qing Dynasty)

• **银镀金臂钏（清）**

Gilt Silver Armlet (Qing Dynasty)

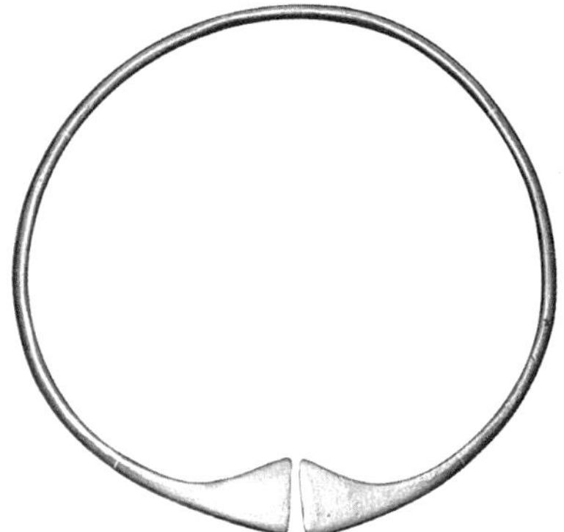

金臂钏（商）
Gold Armlets (Shang Dynasty)

象牙臂钏（清）
Ivory Armlet (Qing Dynasty)

戒指

戒指是套在手指上环状形饰物，古称“指环”。据记载，大约在距今4000多年前的中国就已有人佩戴戒指了。古代女子戴戒指是很普遍的。戒指最初在宫廷中用以记事，也是一种“禁戒”“戒止”的标志。后来，戒指逐渐变成女性的装饰品。

戒指除了装饰和禁戒的作用之外，还可充当婚姻的信物。早在东汉时期，民间就已经出现把指环当作寄情之物的做法，青年男女表达爱慕之情，往往通过赠送指环的方式来实现。今世男女结婚之时，往往有赠送“婚戒”之举，这种风俗就是从古代流传下来的。

Ring

According to the historical record, the ring was worn approximately 4000 years ago. It was very common for imperial women to wear rings, which was used for keeping a note. It was also a symbol of "forbidden". Later it evolved into an ornament for women.

Along with the ornament and "forbidden", the ring served as marriage token. As early as the Eastern Han Dynasty, the ring as a token of love was exchanged between men and women in love. Today the custom of exchanging rings between the bride and the groom was passed down from the ancient times.

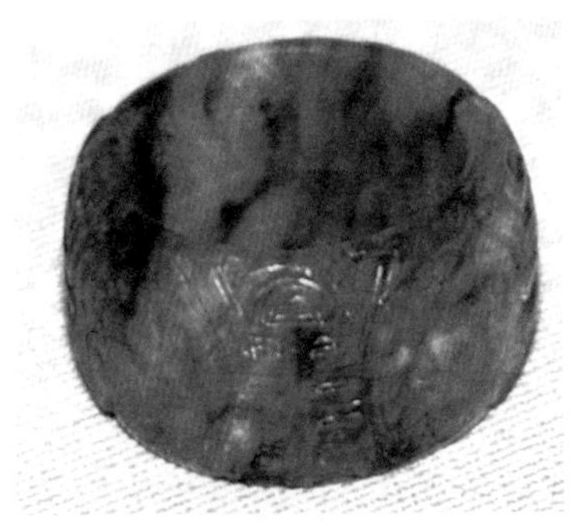

- 翡翠雕蝠寿纹戒指（清）
  Jadite Ring Engraved with Bat Pattern and "Longevity" Character (Qing Dynasty)

- 金錾蝴蝶双喜纹戒指（清）
  Gold Ring Engraved with Butterfly Pattern and "Double Happiness" Character (Qing Dynasty)

- **银梅花纹戒指（清）**

  Silver Ring with Plum Blossom Pattern (Qing Dynasty)

- **银步摇戒指（清）**

  步摇戒指是指挂缀小铃铛的戒指。步摇戒指走一步摇一下，听其声响可以分辨女子走路的姿态是否轻盈。

  Silver *Buyao* Ring (Qing Dynasty)

  Buyao ring referred to the ring attached to small bells. *Buyao* ring would make sound with each step, which could help to judge if her manner of walking was graceful.

• **银珐琅彩戒指（清）**

这三款银珐琅彩戒指造型简洁饱满，加上珐琅彩工艺，显得十分精致。

Silver Rings with Color Enamels (Qing Dynasty)

The shape of these three silver rings tended to be simple but the craftsmanship of color enamels was exquisite.

• **银戒指（清）**

Silver Ring(Qing Dynasty)

护指

清代，皇宫贵妇们流行戴一种镶珠嵌玉的金属或景泰蓝指甲套，称为“护指”，用以保护她们的指甲，同时显示其尊贵的地位。在中国古代，地位高的男女往往会留长指甲，以显示他们无须劳动、地位高的身份。由于指甲很脆弱，人们便发明了这种保护指甲的护具，后来渐渐发展为一种装饰品。清代的护指纹饰精致，雕刻的图案非龙即凤，十分华美。

Fingernail Cover

Ladies in the royal palace of the Qing Dynasty wore fingernail covers which were made of metal or cloisonné. Fingernail covers were used to protect fingernails and show the distinguished social status. In ancient China, both men and women who enjoy high status would have long fingernails to show the fact of freeing from labor work. Fingernail covers were invented to protect the fragile nails but they gradually became the ornament, especially in the Qing Dynasty, such patterns as dragon and phoenix on the fingernails were engraved delicately.

• **金护指（清）**

Gold Fingernail Cover (Qing Dynasty)

扳指

扳指最初是一种护手的工具，用犀牛角、兽骨等制成，在拉弓射箭时戴在勾弦的手指上，用以扣住弓弦，同时防止弓弦擦伤手指。直到17世纪清朝建立后，扳指才逐渐发展为首饰。贵族子弟及附庸风雅的富商巨贾纷纷以贵重材质，如象牙、水晶、玉、翡翠、碧玺等制作扳指，用于相互攀比炫耀。

在清代，佩戴扳指是有严格的等级限制的。翡翠、玛瑙、珊瑚等名贵材料制作的扳指，除了王公贵胄，一般人是不能随意佩戴的。扳指以翡翠质者为首选，其色泽澄浑

Thumb Ring

Thumb ring originally was a kind of protector made of rhino horn and animal bones. When people shot an arrow, it was worn on the finger to draw the bow to avoid getting hurt. It was not used as an accessory until the seventeenth century when the Qing Dynasty was established. To show off their wealth, noblemen and businessmen used precious materials to make thumb rings, such as ivory, crystal, jade, jadite and tourmaline.

In the Qing Dynasty, there was a strict hierarchy for thumb ring wearing. The thumb ring which was made of jade, agate, coral and other valuable materials was exclusively for the offspring of the

- **金双喜字扳指（清）**

此扳指里外层皆由纯金打造，中间夹以木质内胆。外层环绕金圈镂雕五个双喜字，均匀排列，上下边沿各饰上回字纹图案。

Gold Thumb Ring with "Double Happiness" Character (Qing Dynasty)

Wooden interior was embedded in the outer layer of gold. Five "double happiness" characters were hollow-out carved symmetrically on the gold. *Huizi* pattern was engraved on the upper and lower edges.

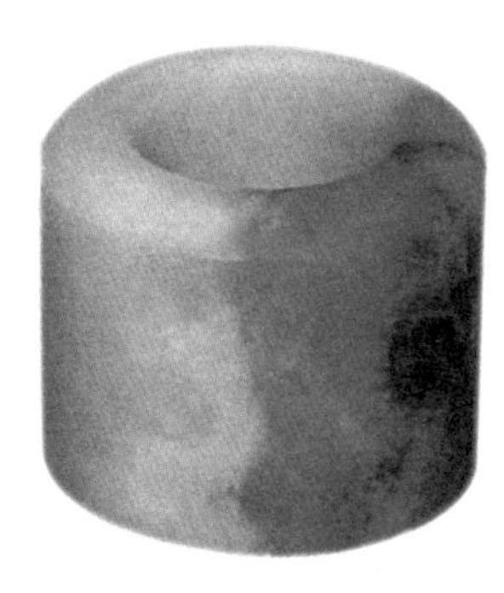

• **翡翠扳指（清）**

贵族扳指以翡翠质者为上选，颜色碧绿而清澈如水者更是价值连城。

Jadite Thumb Ring (Qing Dynasty)

The top thumb ring worn by the nobility was made of jadite. The emerald green thumb ring like clear lake water would be priceless.

• **墨玉扳指（清）**

墨玉通常为黑灰色，多数同青玉相近。像这种纯黑如漆的板指极为罕见。

Black Jade Thumb Ring (Qing Dynasty)

Most of black jade had similar color to gray jade. It was rare to see the color of jade as black as coal.

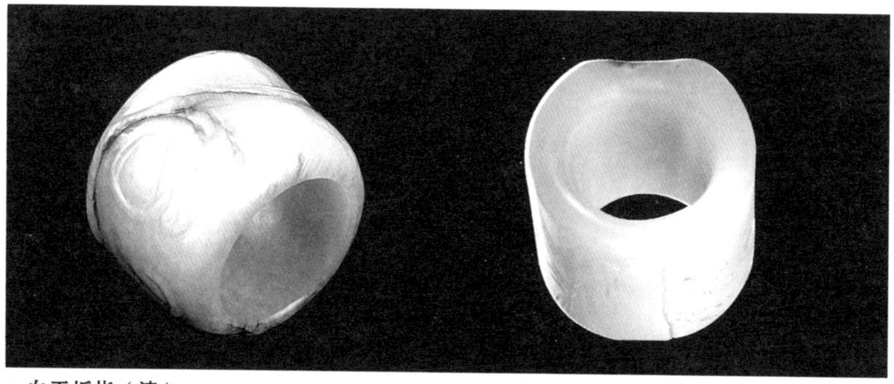

• **白玉扳指（清）**

White Jade Thumb Ring (Qing Dynasty)

• 绿釉瓷扳指（清）
Green-glazed Porcelain Thumb Ring (Qing Dynasty)

• 翡翠镶金里扳指（清）
Emerald Thumb Ring with Gold Inner Ring (Qing Dynasty)

不一，而且花饰斑纹各异。普通的贵族子弟佩戴的扳指以白玉者为多，而平民百姓佩戴的扳指以象牙、瓷质为多。

nobility, princes and dukes, among which Jadite was the first choice because of its mixed colors and diverse natural patterns. The children in well-off family mostly wore jade thumb rings while the materials for common people were mostly ivory or porcelain.